T0093871

Impact of Materials on Society

IMPACT OF MATERIALS ON SOCIETY

First Edition

Eds. Sophia Krzys Acord, Kevin S. Jones, Marsha Bryant, Debra Dauphin-Jones, and Pamela Hupp

With support from the Materials Research Society, Department of Defense, National Science Foundation, and the University of Florida

LIBRARYPRESS@UF
PERRY COLLINS, EDITOR
TRACY MACKAY-RATLIFF, COVER DESIGN
GAINESVILLE, FLORIDA

ufl.pb.unizin.org/imos/

This book was produced with Pressbooks (https://pressbooks.com) and rendered with Prince.

Contents

LIBRARY PRESS@UF

AN IMPRINT OF UF PRESS AND GEORGE A. SMATHERS LIBRARIES

UNIVERSITY *of* FLORIDA

535 Library West • PO Box 117000 • Gainesville, FL 32611-7000
librarypress.domains.uflib.ufl.edu • librarypress@uflib.ufl.edu

ISBN13: 978-1-944455-14-9

Library of Congress Cataloging-in-Publication Data

Acord, Sophia Krzys, editor, author. | Jones, Kevin S., editor, author. | Bryant, Marsha, editor, author. | Dauphin-Jones, Debra, editor. | Hupp, Pamela, editor. | Gillespie, Susan D., author. | Sassaman, Kenneth E., author. | Eaverly, Mary Ann, author. | Curta, Florin, author. | Adams, Sean, author. | George A. Smathers Libraries, publisher.

Impact of materials on society / by Sophia Krzys Acord, Kevin S. Jones, Marsha Bryant, Debra Dauphin-Jones, and Pamela Hupp.

Gainesville, FL : Library Press @ UF, 2021 | Includes bibliographic references | Summary: This textbook supports the Impact of Materials on Society course and teaching materials, developed with the Materials Research Society. The textbook offers an exploration into materials (including ceramics, clay, concrete, glass, metals, and polymers) and the relationship with technologies and social structures. The textbook was developed by an interdisciplinary team from Engineering and Liberal Arts and Sciences, including anthropologists, sociologists, historians, media studies experts, Classicists, and more.

LCSH: Materials science. | Material culture. | Science and the humanities. | Engineering and the humanities. |Materials–Study and teaching.

LCC TA401 / TA492 (ebook)

Acknowledgments

This book serves as a textbook for the Impact of Materials on Society (IMOS) course. The editors and authors wish to acknowledge the support of the University of Florida that made the creation of this textbook possible: especially the Center for the Humanities and the Public Sphere, Department of Materials Science and Engineering, Creative Campus Catalyst Fund, Office of the Provost, George A. Smathers Libraries (especially Perry Collins and Laurie Taylor), University Press of Florida (especially Meredith Babb), and the College of Education. We would also like to acknowledge the Office of Corrosion within the Department of Defense, specifically Dan Dunmire and Greg Redick, for their support and guidance in developing the IMOS course. In addition, we would like to acknowledge the Materials Research Society (MRS) and the many MRS scientists who provided content and input when developing the course, and specifically Richard Souza for believing in the project and his constant encouragement. Finally, we wish to acknowledge the National Science Foundation for supporting the dissemination of the IMOS course and this textbook within the United States, as well as internationally.

Understanding the Material World

SOPHIA KRZYS ACORD AND KEVIN S. JONES

Innovation often happens at an interface. In materials science and engineering (MSE), groundbreaking discoveries have occurred at the interfaces of two or more different materials. For example, the modern computer chip relies extensively on the properties exhibited where metals meet semiconductors. Similarly, innovative thinking can happen at the interface of multiple academic disciplines. To revisit our example, the modern computer chip operates at the intersection between technological devices, the personal lives of those who create and use these devices, and the communities in which they live. This textbook brings the humanities and humanistic social sciences into dialogue with MSE to explore the synergies between human life and materials innovation in societies ranging from pre-civilization to the present. It is only in dialogue between the properties of the "stuff we make" (the domain of MSE) and the sociocultural forces that shape and are impacted by this "stuff" (the terrain of the humanities), that we can fully understand our material world.

The need for close dialogue between the humanities and sciences has been echoed since the origins of the academy. As stated in an article in *The Chronicle of Higher Education*, "An educational system that merges humanities and sciences, creating whole-brain engineers and scientifically inspired humanists, fosters more than just innovation. It yields more-flexible individuals who adapt to unanticipated changes as the world evolves unpredictably."[1]

This is particularly important today, as we increasingly see science not only as the pursuit of natural truths but also as a portfolio of solutions to national and global problems. In addressing these problems, it is important for scientists and researchers to proceed intentionally and with comprehensive knowledge of the social and cultural worlds in which these problems are manifest. Two federal funding agencies, the National Science Foundation (NSF) and National Endowment for the Humanities (NEH) were established in the United States in 1950 and 1965, respectively. In the report leading to NEH's founding, the authors wrote, "If the interdependence of science and the humanities were more generally understood, men would be more likely to become masters of their technology and not its unthinking servants."[2] The report's authors, hailing from leading universities as well as the US Atomic Energy Commission, IBM Corporation, and New York Life Insurance, knew that connecting the humanities and sciences would help us make informed judgments about our control of nature, ourselves, and our destiny.

The 1980 Commission on the Humanities again revisited the relationship of the humanities and sciences, noting that "social and ethical questions are intrinsic to science and technology. In these respects, science and technology have been a domain of the humanities in Western culture since its Greek origins."[3] The same can be said of the close relationship of science and technology to the humanities in all cultures because cultural beliefs and social needs guide how humans fabricate and adopt materials in any context. The humanities and so-called STEM (science, technology, engineering, and mathematics) fields are allies in world making. Advances in science and engineering are themselves social products, expressions of the cultural drives of a society. Furthermore, as the disciplines that seek to understand the values and choices of human decision-makers, the humanities can "awaken scientists and technicians to problems of which they may not have been aware." [4] These problems may involve questions as to the ethical dimensions of technical innovation or possible unintended consequences, but they can also be human needs addressed through technical innovation. As noted more recently, in the 2013 *Heart of the Matter* report produced by the American Academy of Arts and Sciences Commission on the Humanities and Social Sciences, all disciplines must come together in addressing the grandest challenges of our time, including the provision of clean air and water, food, health, energy, education, and human rights and safety.[5]

There is a growing recognition of the importance of holistic thinking among scientific agencies as well. The National Academy of Engineering has categorically stated that today's engineers need to be more than individuals who simply like math and science. They must be "creative problem-solvers" who help "shape our future" by improving our "health, happiness, and safety."[6]

And in 2012, the Accreditation Board for Engineering and Technology (ABET) added two new criteria for engineering education that emphasize the social ends of engineering work.[7] These criteria were modified in 2017 to the following:

1. (2) an ability to apply engineering design to produce solutions that meet specified needs with consideration of public health, safety, and welfare, as well as global, cultural, social, environmental, and economic factors
2. (4) an ability to recognize ethical and professional responsibilities in engineering situations and make informed judgments, which must consider the impact of engineering solutions in global, economic, environmental, and societal contexts

Effectively, these ABET criteria recognize that engineering is truly multidisciplinary. Engineering does not just engage our scientific knowledge of the physical world; it also engages social and cultural awareness of the world in which engineers (and the eventual consumers of their work) live.

Unfortunately, the need for social and scientific dialog in technological innovation is not being met by coursework at most colleges and universities. A 2014 report published by the American Academy of Arts and Sciences discovered that humanities and STEM majors largely dwell in different silos during their educational pathways.[8] This is an impediment to addressing problems that require working with people from other disciplines, as well as recognizing the importance of diverse backgrounds within individual disciplines. Prior research has shown that discussing case studies of how technical products affect people in their daily lives is one demonstrated way to meet the ABET criterion of "understand the impact of engineering solutions in a global. . . societal context."[9] Moreover, existing studies demonstrate that teaching science through the lens of social context and issues contributes to students having more nuanced understandings of engineering solutions as well as a more interdisciplinary and problem-centered conception of scientific inquiry.[10]

The Impact of Materials on Society

This textbook is intended to accompany a freshman-level undergraduate course created by University of Florida faculty with support from the Materials Research Society, Department of Defense, and National Science Foundation. The course examines the discovery, development, and use of materials over time in order to distill lessons from the past that may guide materials engineering innovation in the future. Developed by a team of materials engineers and humanities scholars, this textbook operates at the intersection of material culture and materials science and is intended for use by students of both engineering and the humanities. The stories in this textbook may also accompany a range of academic initiatives including high school coursework, museum exhibits, and lifelong learning programs.

If the goals of these authors are to create curricular dialogue between the humanistic social sciences and engineering, why choose the terrain of MSE to begin? Put simply, everything in our lives is made from a material. MSE is possibly the most ubiquitous form of engineering that we encounter on a daily basis. We have even named societal epochs after materials (e.g., the Stone Age, the Iron Age, etc.) Moreover, the basic properties of materials (e.g., ductility, melting point, density) are intuitive, requiring little advanced mathematics, chemistry, or physics knowledge. The artifacts that we make with materials—from skyscrapers to quarters to microchips—are shaped into cultural currency in our world. Writing about the unfortunate mistrust between the sciences and humanities in 1958, British botanist Sir Eric Ashby noted that technology could be "the cement between science and humanism" because technology is concerned not with science per say, but rather with the "application of science to the needs of society."[11]

MSE as a field has grown out of our desire to make things that better our lives; there has always been an interest in the stuff we use to make these things. From manipulating materials existing in nature like wood, bone, stone, and clay, to creating new materials through the smelting of metals or firing of clays to create ceramics, civilization evolved alongside our ability to manipulate materials. For millennia, people have learned through trial and error how to manipulate existing materials to create new materials with new properties. They have handed this knowledge down through generations. In some cases new materials were discovered by alchemists seeking to create gold. As the science of chemistry evolved, we began to create completely synthetic materials such as polymers, and eventually we developed the tools to examine the microstructure. This led to the ability to correlate how processing the material affected its structure and how that in turn affected the properties of that material. Understanding how this relationship influenced the performance of a material led to the birth of the discipline of MSE in the middle of the twentieth century. By combining earlier fields like ceramics and metallurgy with MSE, our understanding of materials continues to evolve rapidly with novel innovation occurring at the interface of the study of different materials.

But it would be foolish to end a discussion of a material's properties at what is physically discernable. After all, materials innovation does not happen in a vacuum. Materials are themselves caught up in social regimes of power and value. A diamond may be "a girl's best friend," as Carol Channing and Marilyn Monroe once crooned, but diamonds have also been mined by children or sold to finance wars or used to create highly conductive thin films with potential applications in microelectronics. A material also has a lifespan, which begins at its mining or synthesis and fabrication and continues through its recycling or disposal; at each stage it is involved with a network of people, technologies, and other materials. A material can have social properties, such as health, environmental impact, economic value, and social status. And, as demonstrated here in the chapter on plastics, a material can mean different things to different people. This study of how people imagine, fabricate, understand, and use the things they need in community with others is the basic terrain of the humanities and

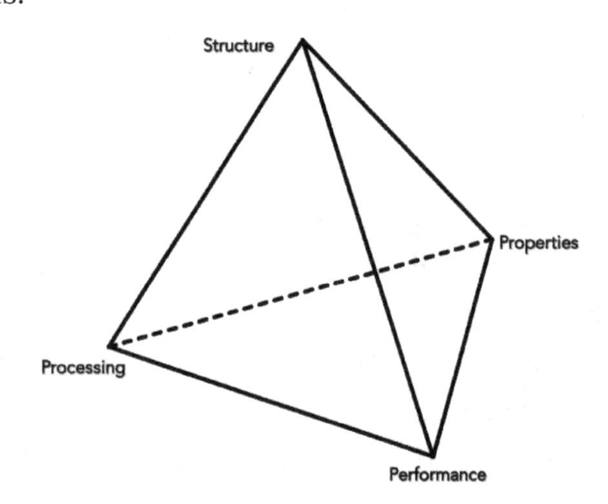

Figure 1.1 Engineering perspective on the materials tetrahedron.

humanistic social sciences. This socio-cultural literacy can be as important to engineering innovations as technical literacy.

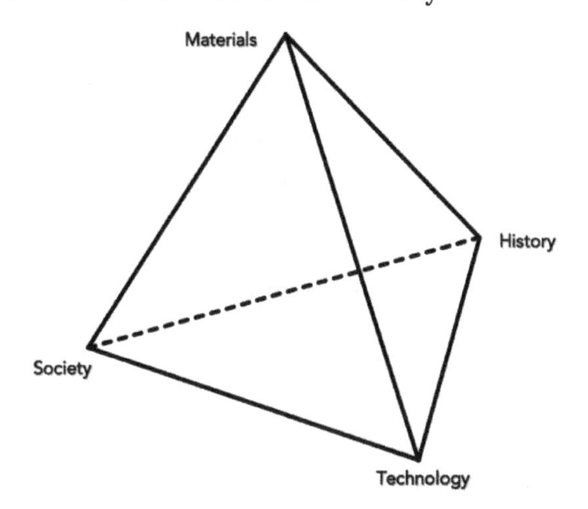

Figure 1.2 Sociocultural perspective on the materials tetrahedron.

In creating this textbook, we have been inspired by other texts exploring the products of materials engineering at use in the world, such as *The Substance of Civilization* (1998) and *Stuff Matters* (2014), both written by materials engineers.[12] In its focus on the sociocultural values that shape our use of materials, however, this textbook is more closely inspired by *The Social Life of Materials* (2015).[13] Its collaborative ethos echoes an earlier anthology edited by University of Florida engineering and classics faculty entitled *Engineering and Humanities* (1982), which represented an effort to share the lessons of the liberal arts with engineers to encourage them to entertain ambiguity and complexity rather than straightforward solutions.[14] We are also indebted to work in the interdisciplinary area of science and technology studies, which unites the study of scientific practices with the study of the use and social impacts of technological artifacts.

This book is unique in that it explicitly aims to generate social concepts from different humanities (and humanistic social science) disciplines that create ways to examine materials from different perspectives. This book is not "humanities for engineers" or "engineering for humanities"; rather, it is an attempt to show how humanists and engineers can improve interdisciplinary communications through innovations afforded by materials. The chapters examine the social contexts of the kinds of problems material scientists are challenged to solve, and the social-political and material consequences of their solutions, many of which are unintended. In so doing, it produces a *usable history*, or distills lessons from the humanities that can guide the development of future engineering solutions and anticipate their social impacts, thus allowing us to prepare for them thoughtfully and intentionally. By sharing lessons from a variety of different disciplines for the study of materials, this text gives students the tools to connect the fundamental concepts and methodologies that they are learning across the curriculum to help address future challenges.

Enduring Knowledge Statements

Taken together, these chapters provide case studies that support five big ideas necessary to understand and predict the impact of materials on society. We refer to these as "Enduring Knowledge Statements":

1. **Materials shape the human experience, and vice versa.** Just as the technologies created through the discovery and use of new materials have extended possibilities for human action, materials are themselves dependent upon humans to create and sustain them. (As we make things, things make us.) Increasingly, this is quite literal as materials are extending our bodies through tissue engineering, polymer scaffolding, even nanomaterials for everything from drug delivery to sensory augmentation. This relationship of mutual dependency between materials and society is further elaborated through the principle of "entanglement", discussed in the chapter on clay.
2. **Materials can be manipulated to solve technical and sociocultural problems.** Material scientists and engineers have played important roles in civilizations throughout human history, even if they have not always looked like today's engineers in white lab coats or "bunny suits." In addressing technical problems such as building sustainable shelters, creating durable tools, and controlling matter at the nanoscale, materials engineers are also developing technologies that are put to use to address basic human needs for shelter, food, healthcare, security, and communication around the world. As a result, MSE is involved directly in addressing social, cultural, economic, political, and ethical challenges as well as technical challenges.
3. **Materials have intrinsic physical properties, only some of which are selected as more relevant in shaping society based on cultural perspectives. The use of a material can change.** It is important to dispel the notion in engineering that "if we build it, they will come." The history of materials engineering is littered with examples of discoveries and inventions that were never adopted in society because they were not well suited to their society's social needs and cultural values. (The chapters on aluminum and polymers in this textbook tell two such stories.) Moreover, engineers are themselves members of the societies in which they work, and unavoidably hold some of the cultural biases and social assumptions of their non-engineering peers. These assumptions can shape the ways in which engineers decide to manipulate materials in a particular time and place. Engineers need to understand how social and cultural forces shape their own work and the reception of their inventions in order to think broadly and creatively.
4. **The impact of materials on society varies with the cultural and historical context**. Higher education tends to teach the same set of values that are shared by the scientific community; people who disagree with science, particularly on religious grounds, do not disagree about the facts, they disagree with the values that underpin

certain scientific claims.[15] In this context material impacts can be negative.

5. **Understanding the impact of materials on society requires a variety of approaches in the humanities, social sciences, and sciences.** As is already evident, the impact of materials on society can be understood only by examining the many intersections of materials and their properties, the technologies they have enabled, social needs, and cultural values. It is important not only to look at what we do with materials-based technologies, but also how our own mindsets shape how we do and do not choose to manipulate these materials in the first place. In taking this holistic approach to understanding materials in our world, we can also combine expertise from multiple disciplines to build more sustainable technologies. For example, we may be able to make more efficient solar cells using cadmium, but if we consider the negative impacts associated with cadmium's toxicity from the beginning, we can insist upon making the cells with an alternative material and innovate in that direction. Thinking broadly ensures that we do not build something with negative social consequences.

This in-depth case study approach to examining materials in past societies also raises several other themes for study across these chapters. For example, looking historically at materials raises the point that materials have a temporality: some last longer than others. Do we want to plan for the obsolescence of materials when we first create and begin to use them? More and more, we are taking a cradle-to-cradle approach in the lifecycle of a material, meaning we are moving away from the use-and-discard approach. To take another example, history tells us that materials innovations often come from unpredictable places, and that contemporary researchers can still learn quite a bit from how earlier humans manipulated and used materials (for example, Roman concrete). The chapters on steel and plastics also note the undercredited roles of some entrepreneurs and their home experiments. Finally, humanistic inquiry can point out critical issues related to the use of materials. Our applications of materials discoveries to social needs are seldom, if ever, neutral. Materials can be used in ways that reinforce disparities based on class, gender, race, and ability, as well as power imbalances between societies. Materials can also be used in ways that impede the health and security of certain communities involved in their production and consumption. Rather than see the humanities as critics of the sciences, however, we must remember that skepticism is itself part of the scientific process.

Summary of textbook chapters

This textbook is designed to be used as part of the Impact of Materials on Society curriculum, hosted on the LibraryPress@UF and Materials Research Society websites.

The curriculum is made up of weekly modules that examine different materials and their impacts on society, from human prehistory to the future. Each week discusses a different material or class of materials. It begins with an introductory lesson as to the material properties, processing and structure of a material from a MSE perspective, followed by a historical case study of that material that introduces a social principle, a video of future material innovations, and finally an activity that links the study of the past and its social principle to the material of the future.

Each of these chapters provides an overview of the social use of a particular material. Some of the chapters focus on a particular place and time in human history, such as 1960s America, the Neolithic settlement Çatalhöyük, or ancient Rome. Other chapters use cross-cultural comparison to study the impacts of materials beyond individual societies, such as the development of inter-society trade networks ushered in with bronze, or the ways in which writing materials contributed to different forms of political power within ancient, medieval, and pre-modern societies. In addition to articulating the ways in which this material was enabling to society at particular points in time, each chapter also provides a lesson on the impact social and cultural systems played in the development of materials. Taken together, these social lessons (e.g., entanglement, creative destruction, etc.) provide a toolkit for future researchers and scientists to design materials more intentionally to address human needs, anticipate and prevent associated pitfalls, and appreciate the role of collaborative inquiry in innovating future materials.

This book is in no way exhaustive of the lessons that can be taken from the history of materials, and from the many disciplines that make up the humanities and social sciences, to examine the impact of materials on society. The current chapters draw lessons from cultures around the globe, with a focus on the history of societies within present-day Europe and North America. This is a living textbook that invites diverse contributions of all human cultures. We hope that the modular template of these chapters will inspire other humanities instructors to collaborate with materials scientists and engineers to expand and add further case studies from all corners of the globe to this open textbook.

It is also important to discuss the organization of this textbook. Certainly, our readers will be familiar with the common practice of applying a chronological order to the human use of materials, (e.g., moving from the "ages" of stone, to bronze, to silicon). More recent scholarship, however, has demonstrated that this framework is an instrument of colonialism in that it seeks to privilege the use of certain materials as more advanced than others. It also implies that the use of some materials replaces others, which is incorrect. After all, in our current "age of silicon," we still create stone tombstones. The use of some materials is more appropriate to certain contexts than others. An indigenous society today may find that stone tools remain completely appropriate for their uses; this does not mean that their society is somehow stuck in the past. Materials do not follow a historical chronology. So, while the historical case studies presented in this textbook loosely follow a chronological order, each chapter stands on its own and chapters may be read in a different order or adapted as needed to learn about the past and ongoing impacts of different materials.

Although this textbook seeks to eschew the idea of a chronological, sequential order of materials use throughout time, the case study focus of each chapter on a different material at a different point in human history does underline another important theme in this textbook: materials that we may neglect or consider ordinary today have actually had major multifaceted impacts that continue to be felt in contemporary societies. For example, while most of our readers will be familiar with the better-known "revolutions" such as the "Neolithic Revolution" that ushered in the transition of hunter-gatherer societies to agriculture, the "Industrial Revolution" and role of iron and steel, and the "Information Revolution" and role of silicon, there are lesser-known but equally significant revolutions that we must consider alongside these. Other chapters examine the "Soils Revolution," the "Concrete Revolution," and the "Polymers Revolution."

"Clay: The Entanglement of Earth in the Age of Clay" by Susan Gillespie introduces us to the widespread social significance of clay, effectively the "steel of early man." In an archaeological investigation of Çatalhöyük, an important Neolithic settlement (7400 BCE–5200 BCE) in then Mesopotamia and present-day Turkey, Gillespie reveals the many ways in which humans at this time depended on clay for shelter, food, community, and family identity. By investigating these dependencies, we see that clay is also dependent upon humans to manipulate, strengthen, and source it. Gillespie introduces the foundational notion of "entanglement" (originally coined by anthropologist Ian Hodder) as a way of understanding the interdependent relationships between humans and materials that shape how we selectively use materials. The concept of entanglement also illustrates how humans become so comfortable with a particular material that adopting alternative materials is challenging.

"Ceramics: Firing Clay and Flaking Stone" by Ken Sassaman builds on this earthy work to examine early ceramics including glass-like rocks such as obsidian and flint. These adaptable materials can be used in a wide variety of ways. By looking at human transformations of ceramics ranging from around 10,000–13,000 years ago, Sassaman examines how ancestral humans developed the evolutionary skills to manipulate materials with greater and greater precision. Focusing on the particular example of the spear-point, Sassaman introduces the anthropological concept of "operational sequence" to study how the ways in which humans interact with materials reveal new potential uses for these materials even as they cordon off other uses. In addition, by comparing the energy used to manufacture materials with the energy they can generate, this chapter suggests ways in which future ceramics can usher in a more energy-efficient world.

"Concrete: Engineering Society through Social Spaces" by Mary Ann Eaverly introduces us to the little known "Concrete Revolution" through an in-depth look at ancient Rome (800 BCE–400 CE). Complementing Gillespie's discussion of Mesopotamian wall construction, Eaverly examines how, as the first human-made structural material, concrete was used by ancient Romans to create monumental public spaces such as the Coliseum, public baths, and a major series of aqueducts stretching across the Roman Empire. But as Eaverly describes, the ways in which Romans used this material directly reinforced and reflected their ideas about social status and imperial power. By comparing the cultural values of ancient Roman leaders to societal ideas today, Eaverly asks us to consider how our own limited cultural perspectives may shape how we think about using material advances in the future.

"Copper and Bronze: The Far-Reaching Consequences of Metallurgy" by Florin Curta takes a unique look at the material and alloy that ushered in the so-called "Bronze Age" (3000 BCE–1000 BCE). Scholarship generally ascribes the significant social transformations often associated with this time—in agriculture, militarization, and learning—to the technological innovations associated with smelting and casting. Indeed, we might refer to copper and bronze as the first materials refined by humans. However, Curta also highlights the significant roles of trade and political complexity that gave birth to metallurgy. Seen in this way, Curta demonstrates that the advance of materials is due as much to the development of expertise of early engineers and the circulation of their know-how as it is due to the availability of the materials themselves.

"Gold and Silver: Precious Metals and Coinage," also by Florin Curta, examines the first uses of precious metals to create widespread systems of exchange within and across societies. By examining the development of currency systems in complex societies around the world, Curta explains how the unique material properties of gold and silver facilitated the development of coinage systems that citizens trusted to hold value. To explain how a society can give particular value to a material, Curta introduces the important distinction between intrinsic and extrinsic value, or the value of the metal itself versus the value that we give to the metal by inserting it into a political, economic, or social system. In this way, this chapter reveals how societies can, quite literally, take something made of one metal (such as nickel) and give it the same worth as gold.

"Steel: Carnegie and Creative Destruction" by Sean Adams examines the Industrial Revolutions made possible by significant processing advances of iron and steel in the 19th-century United States. Unlike the previous chapters, which examined the large-scale context of materials innovation, this chapter focuses our gaze on the important role of particular individuals in spurring this innovation. In particular, Adams introduces us to Andrew Carnegie's aggressive model of entrepreneurship which reorganized factories, supply chains, and market relationships to build one of the largest American companies of the time: Carnegie Steel. Adams also introduces the concept of "creative destruction" to reveal the double-edged sword of materials innovation; the creation of one material-social system can result in the destruction of another one with associated repercussions for workers and competition. This chapter asks us to consider how future materials creations can recognize the industries they may be destroying and avoid social and economic pitfalls.

"Aluminum: Alcoa and Antitrust," also by Sean Adams, tells another 19th- and 20th-century American story, this time of aluminum. Aluminum, a metal enabled by electricity, was a modern material in search of an application. As one American company, Alcoa, slowly discovered marketable applications of aluminum, it also grew in size and scope to control a majority of the world's aluminum supply. This represents a unique case study to understand federal policy related to antitrust issues and competition in the marketplace, and asks readers to consider whether materials monopolies are harmful to society. As future materials research and manipulation will create more materials without immediate application, this question is important to grapple with to prepare an effective business environment.

"Plastics: Fantastic Plastics in Postwar America" by Marsha Bryant examines the modern "Polymers Revolution" in postwar, mid-20th-century America. As synthetic polymers produced in modern research laboratories, plastics have made an indelible mark on popular consumer culture. Bryant outlines the story of Earl Tupper's invention of Poly-T, an ingenious and flexible new material that was only made popular by the marketing genius of Brownie Wise. Building on other chapters that demonstrate how cultural systems shape our perceptions of the potential applications of materials, Bryant introduces the idea of materials marketing to show how materials acquire (and can be given) particular meanings that shape the ways we publicize and use them. As with the story of Andrew Carnegie, this chapter also emphasizes that a material is not simply made by its creators; its sociocultural impact is powerfully shaped by mediators and entrepreneurs, and factors such as aesthetics are often critical to its success.

"Writing Materials: The Politics and Preservation of Knowledge" by Bonnie Effros surveys a variety of writing materials, including papyrus, parchment, and paper, from ancient Egypt in 2500 BCE to Medieval Europe in 1400 CE. Together, writing materials are significant in shaping patterns of human cognition and building communities of individuals. But the material properties and shelf lives of these "information storage media" suited different social uses ranging from government records, to biblical study, to information sharing and community-building among early nation states. These materials also reveal how the ways in which societies store and exchange information can reinforce hierarchies of wealth and privilege. In applying this lesson to the digital age, Effros asks us to think critically about whether future expansive materials for mass information storage will truly democratize our societies, or exacerbate our differences.

We hope that our readers will also notice how many discussions of materials and their impact span across chapters. Those with an interest in the social impacts of building materials, for example, will find much of interest in the chapters on clay, concrete, and steel. (Indeed, even a materials application as niche as stadium roofing is discussed in both the concrete and plastics chapters.) Those with an interest in art and commemoration will find it interesting to examine the various materials that we've used to express ourselves and what we stand for as a society, from the use of concrete to build monumental structures to glorify the Roman Empire, to the US government's use of the very new aluminum to top the Washington Monument and symbolize its country's innovation.

Comparing these case studies also reveals some larger lessons about the social role of materials. For example, although clay in Çatalhöyük and plastics in modern America seem like worlds apart, both materials were used in ways that reinforced a gendered division of labor in and for the home. In addition, these chapters show the importance of considering the many mediators involved in between materials and society: those that extract, process, transport, market, and even regulate material science and industry. And readers interested in knowing more about how technological innovation happens can pay particular note to discussions of the economies of scale in the writing material, steel, and aluminum chapters. Although a more sobering note is offered by the steel chapter about the possibilities of humans being replaced by machines, the lessons of creative destruction ask us to anticipate and address this issue.

Taken together, the contributions in this volume, and those potentially added in the future, aim to show how expertise from multiple disciplines is necessary to create a robust and effective engineering workforce. We hope these case studies will not only teach our readers facts about the past and present, but also generate questions that we must ask of our societies and engineering and technology industries looking into the future. We must discuss the kind of world that we want to live in, and imagine the ways in which new materials will enable us to build the technologies that inhabit this future.[16] To do this, we need to move away from our opinions about materials, to understanding the facts of their development and the values of the societies that embrace them. Just as materials impact societies, societies shape how materials are employed. Only by understanding all of the physical, cultural, and social elements that make up a society can we act in an informed way.

Notes

1. Julio M. Ottino and Gary Saul Morson, "Building a Bridge Between Engineering and the Humanities," *Chronicle of Higher Education*, February 14, 2016, https://www.chronicle.com/article/building-a-bridge-between-engineering-and-the-humanities/.
2. Commission on the Humanities, *Report of The Commission on the Humanities* (New York: American Council of Learned Societies, 1964), 3, https://www.acls.org/uploadedFiles/Publications/NEH/1964_Commission_on_the_Humanities.pdf.
3. Commission on the Humanities, *The Humanities in American Life: Report of the Commission on the Humanities* (Berkeley: Univ. of Calif. Press, 1980), 13, http://ark.cdlib.org/ark:/13030/ft8j49p1jc/.
4. Commission on the Humanities, *The Humanities in American Life*, 18.
5. Commission on the Humanities and Social Sciences, *The Heart of the Matter: The Humanities and Social Sciences for a Vibrant, Competitive, and Secure Nation* (Cambridge, MA: American Academy of Arts & Sciences, 2013), https://www.amacad.org/multimedia/pdfs/HeartOfTheMatter_AroundTheCountry.pdf.
6. National Research Council, "Changing the Conversation: Messages for Improving Public Understanding of Engineering (Washington, DC: The National Academies Press, 2008), https://doi.org/10.17226/12187.
7. ABET Engineering Accreditation Commission, *Criteria for Accrediting Engineering Programs: 2013-2014* (Baltimore, MD: Accreditation Board for Engineering and Technology, 2012), https://web.archive.org/web/20130118205535/http://www.abet.org/DisplayTemplates/DocsHandbook.aspx?id=3149.
8. Norman M. Bradburn and John G. Hildebrand, *Enclosed in a College Major? Variations in Course-Taking among the Fields* (Cambridge, MA: American Academy of Arts & Sciences, 2014), https://www.amacad.org/content/research/dataForumEssay.aspx?i=1571.
9. National Research Council, "Educating the Engineer of 2020: Adapting Engineering Education to the New Century," National Academies Press: Washington, D.C. 2005, https://doi.org/10.17226/11338.; Nicole DeJong Okamoto, Jinny Rhee, and Nikos J. Mourtos, "Educating Students to Understand the Impact of Engineering Solutions in a Global/Societal Context," in 8th UICEE Annual Conference on Engineering Education. Kingston, Jamaica, 2005.
10. John Bryant, La Velle, and Linda Baggott, "A Bioethics Course for Biology and Science Education Students," *Journal of Biological Education* 37, no. 2 (2003): 91–95, https://doi.org/10.1080/00219266.2003.9655858.; Jennifer L. Eastwood, Troy D. Sadler, Robert D. Sherwood, and Whitney M. Schlegel, "Students' Participation in an Interdisciplinary, Socioscientific Issues Based Undergraduate Human Biology Major and Their Understanding of Scientific Inquiry," *Research in Science Education* 43, no. 3 (2013): 1051–78, https://doi.org/10.1007/s11165-012-9298-x.; Zuway R. Hong, Huann-Shyang Lin, and Frances P. Lawrenz, "Effects of an Integrated Science and Societal Implication Intervention on Promoting Adolescents' Positive Thinking and Emotional Perceptions in Learning Science," *International Journal of Science Education* 34, no. 3 (2012): 329–52, https://doi.org/10.1080/09500693.2011.623727.; Jonathan Stolk and Katherine C. Chen, "Creating a Project-Based Curriculum in Materials Engineering," *Journal of Materials Education* 31, no. 1–2 (2009): 37–44.
11. Eric Ashby, *Technology and the Academics: An Essay on Universities and the Scientific Revolution* (London: Macmillan, 1958), http://www.worldcat.org/oclc/727427138. Quoted in James A. Kent,

"The Role of the Humanities and Social Sciences in Technological Education," *Engineering Education* 68 (1978): 725-34.

12. Stephen L. Sass, *The Substance of Civilization: Materials and Human History from the Stone Age to the Age of Silicon* (New York: Arcade Publishing, 1998), http://www.worldcat.org/oclc/855969222.; Mark Miodownik, *Stuff Matters: Exploring the Marvelous Materials That Shape Our Man-Made World* (New York: Houghton Mifflin Harcourt, 2014), http://www.worldcat.org/oclc/855969222.
13. Adam Drazin and Susanne Küchler, eds., *The Social Life of Materials: Studies in Materials and Society* (London: Bloomsbury Publishing, 2015), http://www.worldcat.org/oclc/1159404744.
14. James H. Schaub, Sheila K. Dickison, and M.D. Morris, *Engineering and Humanities* (New York: John Wiley & Sons, 1982), http://www.worldcat.org/oclc/1105242973.
15. John H. Evans, in "Scientific Advances and Their Impact on Society," *Bulletin of the American Academy of Arts and Sciences* (Winter 2016): 46–48, https://web.archive.org/web/20200818004649/https://www.amacad.org/news/scientific-advances-and-their-impact-society.
16. Jameson Wetmore, Ira Bennett, Ali Jackson, and Brad Herring, *Nanotechnology and Society: A Practical Guide to Engaging Museum Visitors in Conversation*, (Nanoscale Informal Science Education (NISE) Network; The Center for Nanotechnology in Society, 2013), https://www.mrs.org/docs/default-source/programs-and-outreach/strange-matter.green-earth/nanotechnology-and-society-a-practical-guide-to-engaging-museum-visitors-in-conversations.pdf.

Clay: The Entanglement of Earth in the Age of Clay

SUSAN D. GILLESPIE

> "There is no such thing as 'stone'; there are many different types of stones with different properties and these stones become different through particular modes of engagement." —Chantal Conneller, *An Archaeology of Materials*[1]

Abstract

This chapter showcases human engagements with the most primal material of all—earth itself—beginning in the Neolithic period (which began ca. 9000 BCE in the Fertile Crescent), when people relied on domesticated plants and animals for their livelihood. The Neolithic has also been called the Age of Clay because clay and soils were critical materials for many aspects of daily life. A case study of an important Neolithic settlement, Çatalhöyük, demonstrates how people and clay became interdependent on each other, resulting in an "entanglement" that influenced human actions and values. The Neolithic entanglement with clay, multiplied countless times all over the globe, led to significant historical changes in human society that still reverberate today. This case study also provides general insights for understanding the relationships between humans and materials. How people engage with the potential and actualized properties of materials in production processes is key to understanding the historical trajectories of the impacts of materials on societies.

Introduction

How have archaeologists, and philosophers before them, made sense of human history? From the beginning of the discipline in the 19th century, archaeologists focused on the different materials manipulated by humans over time. Thus they began to organize the human past in a logical way—a series of progressive stages—that is still influential today. But it is only much more recently that archaeologists and other scientists have begun to investigate the impacts of specific materials on human societies. It is not an exaggeration to say that human-material interactions changed history. People continue to be shaped by relationships with certain materials that began long ago. This chapter examines human interactions with a humble material—earth in the form of clay—starting over 10,000 years ago.

Thomsen's Three-Age System

In 1816, Danish antiquarian Christian Jürgensen Thomsen (1788–1865) faced a major challenge. The Danish Royal Commission for the Preservation and Collection of Antiquities had been amassing collections of ancient artifacts from all over the country to house them in what would become the National Museum of Denmark (Figure 2.1). The commission asked Thomsen to organize the various objects for an exhibition to educate Danes regarding their early history.[2] How could he best make sense of them?

Figure 2.1 Thomsen with visitors in the Museum of Northern Antiquities, Denmark, in an 1848 drawing. [Wikimedia Commons.]

Thomsen decided to organize the artifacts by their raw material, which provided clues to their historical contexts. Nearly 2,000 years earlier, the Roman philosopher Lucretius had speculated that the first humans used stone and wood for implements, and only later developed bronze and then iron (see Curta, "Copper and Bronze"). But this idea had never before been tested.

Thomsen reckoned that ancient people still used stone tools after bronze metal-working appeared, and that they continued to employ bronze artifacts after iron was introduced. But to group objects solely by material was not meaningful to the history of Denmark, because it ignored cultural information about how and when past peoples made and used these objects.

To understand better when objects made of these materials were used, Thomsen focused on artifacts from "closed finds" such as burials and hoards (buried caches of objects). With closed finds he could assume that all the items found together were in use at the same time (Figure 2.2). In this way, he could determine which things were probably contemporaneous and utilized by the same peoples. Thomsen ended up with distinct groupings that, he suggested, formed a sequence in time, proving Lucretius's early idea with material evidence. He proposed a series of "ages" in early Danish history: an initial period with only stone artifacts (Stone Age), a later period with both bronze and stone tools (Bronze Age), and a final Iron Age with objects of iron, bronze, and other materials.

Published in 1836, Thomsen's innovative chronological scheme energized the developing field of archaeology. His Three-Age System of Stone-Bronze-Iron was later applied to all of Europe, Africa, and Asia, although the "metal ages" do not pertain to the Americas, Australia, or Oceania. The "Stone Age" was subsequently divided into sub-periods, the earliest two being the Paleolithic (Old Stone Age) and Mesolithic (Middle Stone Age). Both are characterized by tools made from the forceful removal of chips (flakes) from stone (see Sassaman, "Ceramics"). The subsequent Neolithic period, or New Stone Age, was distinguished by a new technology for grinding and polishing stones to make implements.

Figure 2.2 Engraving of early Bronze Age burial in Britain. Visible in the drawing, this "closed find" includes a ceramic beaker, bronze dagger, and a stone projectile point (just above the dagger in the drawing). [Llewellyn Jewitt, from Grave Mounds and their Contents (London: Groombridge and Sons, 1870), 14. Internet Archive.]

What's Missing in the Three-Age System?

Thomsen based his artifact groupings on hard, durable objects of some value, intentionally buried in graves or hoards. His chronological scheme thus neglected the soft, perishable, non-grave-worthy materials used by Denmark's early inhabitants. Our ancestors used many other "earthy" materials that are not represented in the Three-Age System.

Furthermore, there is no sense of how and why certain materials came to be used by earlier peoples at different points in human history. Nor was Thomsen able to articulate how those materials impacted the development of society in advantageous or disadvantageous ways.

This chapter showcases human engagements with a material neglected by Thomsen and Lucretius even though it is the most primal material of all—earth itself—during the Neolithic period.[3] The Neolithic "soil revolution" provides historical background for a case study of the entanglement with clay experienced by the inhabitants of Çatalhöyük, an ancient settlement in modern-day Turkey.

Entanglement is the key idea introduced in this chapter. It refers to the interdependency between humans and things, based on the properties of the materials that things are made of.[4] Entanglement becomes an entrapment that influences human actions and ideas. The Neolithic entanglement with clay, multiplied countless times all over the globe, led to significant historical changes in human society that still reverberate today.

This case study from the deep past also provides a method for analyzing contemporary and future relationships between humans and the materials they depend on. The final section thus draws out some "material lessons" we can use to better understand the impacts of materials on human society.

The "Soil Revolution" and the "Age of Clay"

Changes in the Neolithic Period

Although Thomsen's Three-Age System is overly simplistic, the division of the Stone Age into earlier (Paleolithic and Mesolithic) and later (Neolithic) components is still useful. However, these terms no longer refer simply to changes in stone tool technology, nor to exact periods of time. Instead, **Neolithic** now designates the shift from a nomadic food-collecting to a settled food-producing way of life dependent on domesticated plants and/ or animals. This gradual transition occurred in different parts of the world at various times, beginning about 9000 BCE in the Fertile Crescent area that stretches from Syria into Iraq and Turkey.

Figure 2.3 Reconstructed Neolithic house interior in Albersdorf, Germany. Note the use of clay or earth for parts of the wall, the hearth/oven, and pottery. [Photo by user Nightflyer (2012), shared under a CC-BY 3.0 Unported License. Wikimedia Commons.]

Being tied to the land to raise crops or to tend livestock required more durable houses and other structures. Furthermore, most Neolithic peoples developed pottery vessels to store, serve, and sometimes cook foods (Figure 2.3). These changes were so substantial, modifying the course of global history, that this transition was dubbed a Neolithic "Revolution."[5]

New Engagements with Soil and Clay

Figure 2.4 The great henge (circular ditch and embankment) at Avebury, England, near Stonehenge. Most of the stones from its stone circles have been removed. [Photo by author (2006).]

Although the shift from mobile to settled village life varied tremendously where it occurred across the world, there were certain commonalities to the experience. One of them was an increase in practices that opened up the earth's surface. Excavations were necessary for new forms of architecture and land modification. These included holes for house posts, pits for storage or garbage, "borrow pits" made when removing soil for other purposes, graves for the dead, ponds for livestock, and ditches for drainage, irrigation, or for ritual spaces, such as the circular ditches and embankments (henges) in Great Britain (Figure 2.4).

Digging also exposed new earthy materials lying below the surface, such as flint, limestone, and clay, which were utilized in new tool and building technologies. In the late Neolithic people began to mine and process copper and gold. This early metal working marks the Chalcolithic (copper-stone) period, precursor to the Bronze Age (see Curta, Copper and Bronze). Archaeologist Julian Thomas described these actions of digging into and mounding up dirt as a new "set of relations of reciprocity with the earth itself" with new methods for transforming earthy materials. [6] Put another way, Neolithic societies were "soil-based societies" because soil (broadly speaking) is a common denominator in all the major changes of this new way of life (Figure 2.5).

Key Concept: What is Clay?

Clay is technically defined as the finest sediments, with grain diameters of less than 1/256th of a millimeter. The high surface area of these plate-like grains makes true clay sticky and plastic. However, archaeologists use a less specific characterization that refers to clay's plasticity, treating clay as any fine-grained sediments capable of being molded to an internally cohesive form. "Clay" deposits typically consist of both true clays and non-clays, both fine-grained and coarser-grained sediments.

Soil was necessary to grow the crops, to pasture the animals, to erect more durable structures, and to make pottery. According to archaeologist Nicole Boivin, a veritable "**Soil Revolution**" occurred that has gone unrecognized because we tend to think of soil as the unchanging stuff beneath our feet.

Figure 2.5 An artist's rendering of an early Neolithic village. [Science City of Kolkata. Photo by Biswarup Ganguly, shared under a CC BY-SA 3.0 Unported License. Wikimedia Commons.]

Figure 2.6 Seated "Mother Goddess" flanked by felines. Clay figurine excavated by James Mellaart at Çatalhöyük in 1961; the head is a restoration. [ca 6000 BCE, Museum of Anatolian Civilizations. Photo by Nevit Dilmen, shared under a CC BY-SA 3.0 Unported License. Wikimedia Commons.]

On the contrary, soils underwent dramatic modifications as they were drawn into Neolithic and later technologies.[7]

The dominant earthy material for Neolithic peoples was **clay.** It was used to make all or parts of houses, pottery, figurines, cooking balls, fishnet weights, jewelry, gaming pieces, and many other artifacts (Figure 2.6). In their everyday lives, people were enclosed in clay, manipulated clay, ate from clay, wore clay, and experienced its different forms in close, personal contexts. In addition, clay objects lasted longer than those made of organic materials such as fiber, wood, or bone, and their longevity gave people a new sense of their own enduring histories.

The remarkable increase in material production of clay objects and structures along with emerging transformative clay technologies motivated archaeologist Mirjana Stevanović to insert an "**Age of Clay**" between the Stone and Bronze Ages.[8]

Properties of Clay

Why was clay so important? A major reason, besides its abundance and ease of acquisition, is that clay exhibits the property of **plasticity** or malleability. Clays are fine-grained sediments that, mixed with the right amount of water, can be formed into a variety of shapes.[9])

Clay was critical to the development of many innovative transformative technologies of the Neolithic period. These transformations were not merely mechanical—as in the polishing of hard stones to make axes to cut trees and grinders to process grains—but were increasingly structural and chemical.

Mixing clay with water created structural changes, allowing it to be formed into figurines, bricks, vessels, jewelry, and other objects. Heating and drying those objects—using the sun or with an oven or hearth—effected structural modifications that made them harder and more durable. The innovation of a high-temperature open or closed kiln for firing dried clay objects produced chemical changes, transforming them into ceramics (Figure 2.7; see Sassaman, "Ceramics"). [10]

Figure 2.7 Pottery in the backyard of a potting household in Atzompa, Oaxaca. Mexico. The pots in the center are drying; those in the lower left corner have already been fired in the kiln (out of the photo, on the left). [Photo by author (1984).]

Disadvantageous Properties

However, Neolithic peoples also had to deal with certain problems of working with clay. Clay is heavy and bulky to move from wherever it is mined from the ground to where it is needed. The potters must then process the mined clay, usually by pounding it into a gritty powder and removing impurities. Adding water to clay makes it heavier still and difficult to maneuver.

Clay objects require water to be formed, but as the water evaporates, they tend to shrink and crack. **Temper** (a non-plastic material) was typically mixed in with the prepared clay to reduce shrinkage. Various materials served as tempers, including plant fibers, sand, volcanic ash, limestone, shell, and even ground-up ceramics. Clay objects are also fragile; they break easily, and once broken, lose most of their value.

Activity: Watch a video

This video (https://www.youtube.com/watch?v=AO0F8y3aNOo) features potters in Botswana's Kgatleng District.

- How do the women prepare the clay they have mined to make it usable?
- How do know how much water to add to the clay?
- Do they shape the clay into forms such as bases or coils before they make a pot? Do they use a potter's wheel?
- What kind of fuel do they use to fire (heat) their pottery?
- How would you describe the kiln they use? Is it what you expected?
- How does pottery-making create opportunities for social interactions among women?

For these reasons, adopting a clay-centered technology meant a loss of mobility. Neolithic peoples were less free to move about because of their accumulating possessions and the desire to be close to both clay and water. At the same time, they required more stable settlements in order to consistently control their agricultural fields or livestock. This growing "investment in place," requiring more permanent settlements, was often accomplished by making more durable residences out of earth and maintaining them across generations.[11]

The Entanglement of Clay at Çatalhöyük: Background

Çatalhöyük was one such long-lived Neolithic settlement. It is also an unusually large site, located in south-central Anatolia, southeast of the modern city of Konya (Figure 2.8) in modern-day Turkey. People lived at Çatalhöyük continuously for over 2,000 years, from 7400–5200 BCE.

Figure 2.8 Map of modern Turkey showing the location of Çatalhöyűk.

Archaeologist Ian Hodder, who has directed excavations at Çatalhöyük since 1993, demonstrated the extent to which its Neolithic inhabitants became entangled with clay throughout their settlement's long history. His analysis helps to explain the impacts of clay over time on this and other ancient societies. It also illustrates aspects of the relationships between people and the materials they use that can be applied to many other materials in the past and present.

Çatalhöyük

Like many Neolithic sites in the region, Çatalhöyük ("forked-mound") is a human-made mound (*höyük* in Turkish) created by the continuous building of clay houses atop one another over many generations (Figure 2.9). Excavations in the late 1950s and 1960s led by archaeologist James Mellaart first brought to light its unusual settlement plan. The town consisted of contiguous multi-room rectangular houses all made of clay, including sun-dried clay bricks (called mudbricks).[12]

As of yet, archaeologists have not discovered a town center or non-residential structures. The settlement seems to be all houses.

Figure 2.9 Satellite photo of Çatalhöyűk. The white areas on the main (east) mound are roofs to protect the north and south excavation areas. Note the modern agricultural fields all around the mounds. [Google Earth.]

Because houses abutted one another, most people entered and exited their dwellings through a hole in the roof. Streets are absent, and residents used the flattish roofs as both outdoor space and to walk about the settlement (Figure 2.10). Families apparently controlled their own houses across generations, even making their own mudbricks, as no two households utilized the same clay materials in the same way for their residences.

Figure 2.10 Excavations under the north shelter in 2010 revealed adjoining houses. There are single walls within a structure, but two adjacent structures each have their own exterior wall (see center of photo). [From excavation, shared under CC-BY-NC-SA 2.0 Generic License. Flickr.]

After a period of 50-100 years, the occupants collapsed the roofs, tore down the house walls to about midway, and filled the remaining cavity to serve as the foundation for a new house atop the old. They did not recycle mudbricks from earlier buildings, so each subsequent construction phase required new bricks. As a result of these multiple independent building decisions, the mound developed unevenly, with roofs of individual houses at different elevations. Ladders were likely used to go up and down over house roofs to get off the mound to tend agricultural fields and livestock (goats, sheep, and cattle).[13] The people of Çatalhöyük made and lived in a dynamic world constructed of clay, which impacted every major aspect of their lives.

Clay and the Founding of Çatalhöyük

Besides being integral to the form and growth of Çatalhöyük, clay was essential to its location. The settlement was established on the Konya Plain, the bed of the drying Ice-Age Lake Konya.[14] The absence of stone and scarcity of trees here meant that clay would be the primary material for building and for many other needed objects. In Neolithic times, this lakebed was a source of multiple kinds of clay sediments: marls (highly calcareous clays); backswamp (alluvial) clays formed from the deposition of sediment in the lake; reddish clays with silt; colluvium that accumulated at the base of the growing mound; and gritty clays.

However, the scarcity of fuel and the enormous number of clay bricks needed for construction also meant that people relied on the sun to dry the bricks rather than firing them in kilns. This "subceramic" technology extended to other clay objects, including figurines, clay balls, and some pottery.

Çatalhöyük's first settlers placed their houses in an area of the lakebed with thicker deposits of backswamp clays, rather than on nearby areas elevated by marl deposits. Mudbricks are heavy, especially when wet. By building directly on this chosen clay source, the inhabitants sought to avoid high transport costs. The decision to locate the initial settlement directly on the low-lying, clay-rich areas rather than on natural rises exposed it to flooding. However, as more houses were stacked upon the earliest one, the resulting mound elevated them above the flood zone (Figure 2.11).

A World Heritage Site

Çatalhöyük has many unusual characteristics and provides significant information on the transition to an agricultural way of life. The main mound was continuously occupied between 7400–6200 BCE and experienced eighteen distinct building levels. At its maximum size, it was twenty-one meters tall and extended over thirteen hectares. At different times approximately 3,500–8,000 people lived here, a population size equivalent to a large town or even a small city. A shorter mound, dubbed Çatalhöyük West, was subsequently occupied from 6200–5200 BCE in the Late Neolithic (Chalcolithic) period. Because of its importance, Çatalhöyük was designated a World Heritage Site.

Çatalhöyük is also well known for its mural art, consisting of paintings and low-relief sculptures on clay walls, as well as clay figurines and unique clay supports for erecting cattle horns (bucrania) in the walls of some houses. The Çatalhöyük Research Project, directed by Prof. Ian Hodder of Stanford University, is a 25-year program of excavation, conservation, interpretation, and presentation of findings.

Figure 2.11 South area excavations in 2014 show multiple levels of houses, with the individual bricks visible in some places. Sandbags help conserve the walls against continued slumping and deterioration. Both the alluvial and colluvial sediments were utilized to make bricks and other clay objects later in the sequence of occupation. [Photo by Jason Quinlan, shared under a CC-BY-NC-SA 2.0 Generic License. Flickr.]

In addition to mudbricks, backswamp clay was the principal material for clay cooking balls and some early pottery. But as residents continued to dig out the backswamp clay, they depleted this resource in their immediate environs, while also exposing other clays and marls underneath. The whitish marls (highly calcareous clays) beneath the backswamp clay had their own uses, especially to make the plaster that covered and protected the mudbrick walls and served as mortar for the mudbricks (Figure 2.12). Thus, at Çatalhöyük there was "no such thing as clay"; many similar materials were differentiated by their specific properties, locations in the landscape, and uses over time by the inhabitants.

Extracting clays changed the local landscape. Digging for clay disrupted the flow of water in what was a wetland environment, changing the drainage and thus vegetation patterns. A rough estimate of the amount of clay needed for Çatalhöyük's residential uses during the life of the settlement is an astonishing 675,000 cubic meters! The borrow pits excavated to obtain clay were regularly flooded, filling with additional water-deposited sediments (alluvium). As the mound grew higher, erosional sediments accumulated at its base (colluvium), mixed with artifacts.

In sum, even as clay shaped the Çatalhöyük community, their collective actions also impacted the clay deposits themselves, changing their composition.

Figure 2.12 Repaired patches of clay mortar preserved finger impressions.[Photo from excavation, shared under a CC-BY-NC-SA 2.0 Generic License. Flickr.]

Hodder's Entanglement Model

Drawing upon his excavations at Çatalhöyük, Ian Hodder devised a model of how humans and materials become dependent upon one another, creating an entanglement. [15]

This model is a highly useful method for

analyzing the impact of materials on society. It is based on four simple-sounding premises:

1. Humans depend on things.
2. Things depend on other things.
3. Things depend on humans.
4. Humans depend on things that depend on humans.

(Note that Hodder takes as axiomatic that humans depend on other humans.) His fourth premise, which builds upon the first three, is the **entanglement**—a human-thing interdependency. Applying these premises one by one, Hodder analyzed how the inhabitants of Çatalhöyük became entangled with clay, and how that entanglement changed their lives and their history.

Activity: Watch a video

Ona Johnson and Karis Eklund made this video, "Welcome to Çatalhöyük," in 2004 (https://www.youtube.com/watch?v=CNZRzKChn84&t=84s) to introduce the site to visitors.

- What is so special or unique about Çatalhöyük?
- What does the video tell you about the original wetland environment of Çatalhöyük?
- How was Çatalhöyük unlike a modern city? Why were houses so important?
- What did you think of the ancient artworks made by the Neolithic inhabitants of Çatalhöyük?

Humans Depend on Things

Regarding "things" made of clay, the first premise is indisputable. The earliest houses were built directly upon thicker deposits of backswamp clay needed for structures. Residents of Çatalhöyük and other Neolithic sites depended on their houses of mudbrick walls; interior hearths, ovens, and benches made of clay and plastered with marl to help make them waterproof and durable; and many other clay objects (Figure 2.13).

Figure 2.13 Overhead view of a house (Building 56) in the South Area. Note the extensive use of clay for all the house features, including benches, partitions, and the hearth (center left). [Photo from excavation (2006), shared under a CC-BY-NC-SA 2.0 Generic License. Flickr.]

As Hodder explained, the residents of Çatalhöyük lived in an intimate, sensory world of clay. Clay dust was ubiquitous and got into their hair, skin, and lungs. It also got into their food because they used heated clay balls to cook their meals, such as animal stews. The dead were buried under the floors (within the earlier filled-in houses), where the clay absorbed the liquid and odors of bodily decay.

Things Depend on Other Things

The second premise is a little more difficult to understand and requires a close examination of the word "**thing**." Although this word is so generic it is difficult to agree on a definition, Hodder relied on the influential insights of German philosopher Martin Heidegger.[16] Heidegger observed that the original meaning of "thing" in Germanic languages (including English) was a "gathering" or "assembly." Things actively gather: they gather their individual properties, other things, processes, people, and places.

There are several ways to understand how clay things gathered or assembled. The clay sediments used to build Çatalhöyük "gathered" (were dependent on) such other things as the hydrogeology of the former Lake Konya and the extraction of the backswamp clay to make the mound. This gathering includes the resulting alluvial clay deposited in the borrow pits and colluvial sediments coalescing at the base of the mound. And all of this movement of clay depended on the actions of people and the force of gravity.

Making a clay object, whether a vessel or a brick, also required gathering component parts, each with its own properties, and correctly assembling them. The parts depended on one another to create the object. Hodder describes this assemblage in detail in making the **paste**, the prepared clay ready to be molded or modeled into an object (Figure 2.14). The clay itself brings its own qualities to the paste: grain size and shape, chemical composition, shrinkage factor, cohesion, any impurities not previously removed, thermal properties, and so forth. A mineral temper (a non-plastic addition) will have similar properties added to the mix.

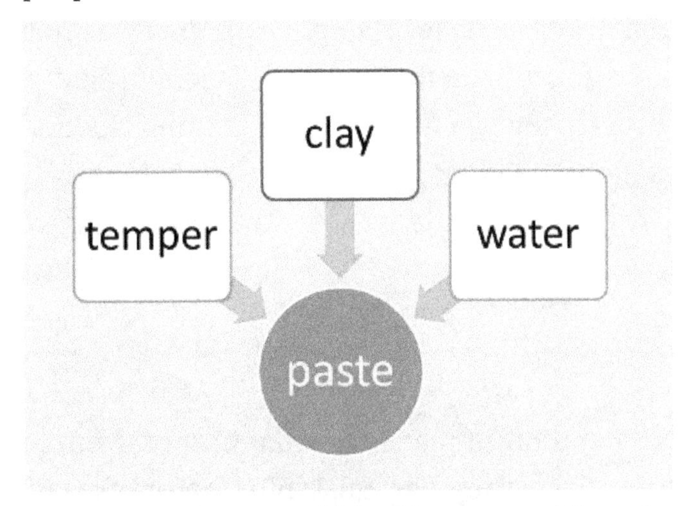

Figure 2.14 Paste as a "thing" is an assembling of the properties of clay, temper, and water. [Adapted by the author from Hodder, Entangled, figure 6.1.]

However, the temper for the early clay artifacts at Çatalhöyük consisted of organic fibers, including wild grasses, straw, and cereal chaff that had to be collected and stored when the grains were harvested.[17] Those activities and materials are part of the gathering process as well. Finally, the water added to mold the clay mix would vary in terms of its proportion to clay and any impurities it might contain. Therefore, making any composite material, such as paste, requires a gathering or assembly.

At a higher scale, a thing is an assemblage because other things, people, and places must come together to manufacture, use, repair, or discard it. As an example of this level of gathering, Hodder diagrams (Figure 2.15) how the marl plaster used to cover the mudbrick walls is a thing.

Plaster assembles the raw material, the marl (calcareous clay), which required certain tools to excavate it from below the backswamp clay deposits and baskets or other containers to transport it to the places for processing. Lime was added to the marl, obtained by acquiring limestone and heating it at a special place (see Eaverly, "Concrete"). Water also had to be carried in containers from its place of origin. Once made, the plaster was applied with certain implements and then burnished to a hard surface with pebbles. This work was dependent on the season of the year and the availability of sufficient laborers. All of these components were "gathered" according to a specific "operational sequence" (see Sassaman, "Ceramics") to make the plaster to maintain the walls, which required frequent resurfacings due to wear and tear. And consider, of course, that the walls were dependent on the plaster to function properly.

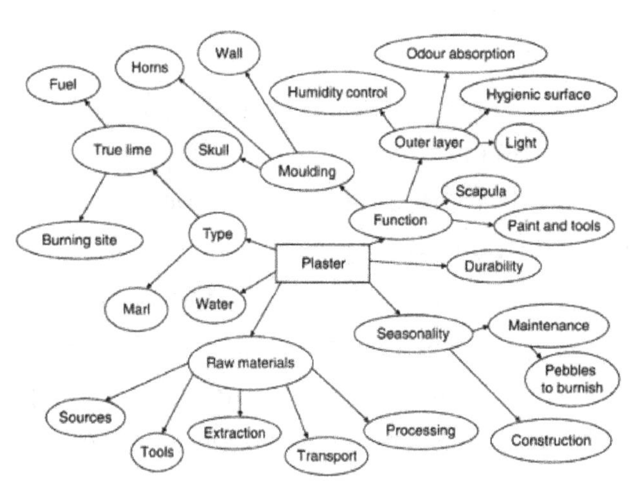

Figure 2.15 This diagram illustrates how clay (marl) plaster (center rectangle) as a "thing" gathers tools, materials, activities, and processes. [From Hodder, Entangled, figure 3.2.]

A more straightforward example of how things depend on other things comes from the houses themselves. Families in two adjoining houses came to build their own house walls next to each other rather than share a common outer wall (see Figure 2.10). This was not due just to a concern for property boundaries. Over time the mudbrick walls would slump or crack, and they depended on the adjacent wall to help them stay upright. Thus Hodder concluded, "All things depend on other things along chains of interdependence in which many other actors are involved. . . . Things in their dependence on other things draw things and people together."[18]

Things Depend on Humans

While things such as mudbricks and pottery depend on humans to come into existence, they also retain that dependency over time because they are unstable. Materials and objects decay, transform, break, fall apart, and sometimes just run out. This dependency is well-illustrated by the mudbrick walls of Çatalhöyük.

As noted, the earliest walls and bricks were made of the backswamp clay. This is a smectitic clay, the name for a category of phyllosilicate minerals that have the high propensity to shrink and swell. In other words, smectitic clays expand quickly when mixed with water, but they shrink a great deal as they dry, and continue to shrink long afterwards. The bricks required tempering with plant material and thick layers of marl mortar to even them out when they were laid because they warped as they dried.[19]

The walls of shrinking mudbrick became more unstable over time, requiring greater human investments to prop them up and keep their surfaces from cracking. As Hodder observed, "The relationships between molecules in the clay produced relationships between people in society at Çatalhöyük as they worked together to solve the problem of collapsing walls."[20] They tried various solutions, including additional layers of plaster and double-walls for mutual support. They also tried to prop up the houses with wooden posts, although this meant reducing the already low numbers of trees in the area.

Over time, larger and heavier bricks were made to create thicker walls, and sandier clays began to be employed. This last change in clay material was likely related to the over-exploitation of the backswamp clays and the use of the sandier or gritty clay deposits underneath them. New technologies were also introduced to manufacture bricks.

All in all, this greater investment of labor and resources in maintaining existing walls and building new houses limited the time and effort that could have been spent on other activities. As Hodder concluded: "People increasingly got trapped by bricks at Çatalhöyük."[21] And the bricks are still a trap today! Keeping the exposed 9,000-year-old mudbrick walls from crumbling is a constant chore for archaeologists and conservators, requiring regular injections of chemicals, consolidants, and grouts.[22]

Entanglement: Things Depend on People Who Depend on Things

Entrapment is the historical consequence of entanglement as a self-propelled spiral of consequences. Hodder represents the entanglement of clay as a **tanglegram**, or interconnected web of dependencies (Figure 2.16). The increasing dependence of Çatalhöyük's residents on the clay for their structures required ever greater investment in maintaining and repairing them. As a result, they changed their activities, their environment (as they dug up the clay), and their social relations. This cascade of consequences well illustrates how an entanglement traps people into doing some things and limits their abilities to engage in alternatives.

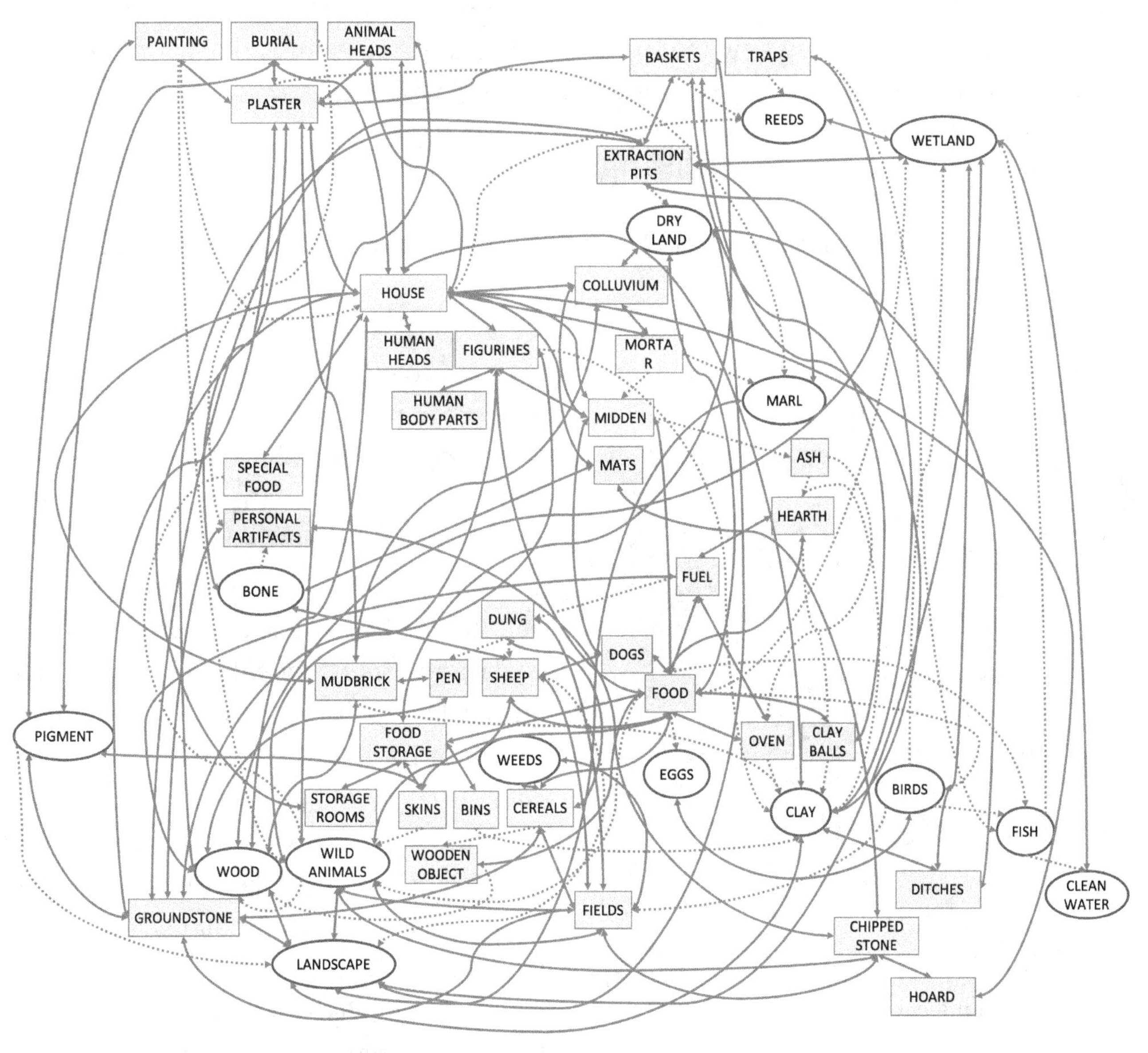

Figure 2.16 A "tanglegram" graphically shows the interdependencies of people and things centered on clay (clay is the oval in the lower right corner, connected to "oven," "hearth," and "mudbrick"). Click on the image to inspect these relationships more closely. [From Hodder, Entangled, figure 9.2.]

Entanglement and Social Change

The entanglement of Çatalhöyük's residents with their mudbrick houses reveals how impossible it would have been for them to change basic architectural materials or construction technology. They were far too invested in their current practices and material dependencies to abandon them.

Entanglement thus provides a framework for understanding how people undergo societal changes—or alternatively, how and why they attempt to prevent change from happening. Hodder's thesis is that the entanglements of humans and things create a historical trajectory that influences the success or failure of specific social and cultural traits. Because of the entrapment caused by one or more entanglements, people are generally unable to adopt a new material or technology, or cannot realize its benefits, unless it fits into an existing technology and labor regime.

From Cooking Balls to Cooking Pots

A good example of this latter scenario from Hodder's case study is the gradual shift from clay cooking balls to cooking pots.[23]

Archaeologists uncovered massive numbers of clay balls from the lower, earlier levels of occupation (Figure 2.17). Many of them were likely used to cook food, as this is a common technology found at equivalent time periods elsewhere in the world.

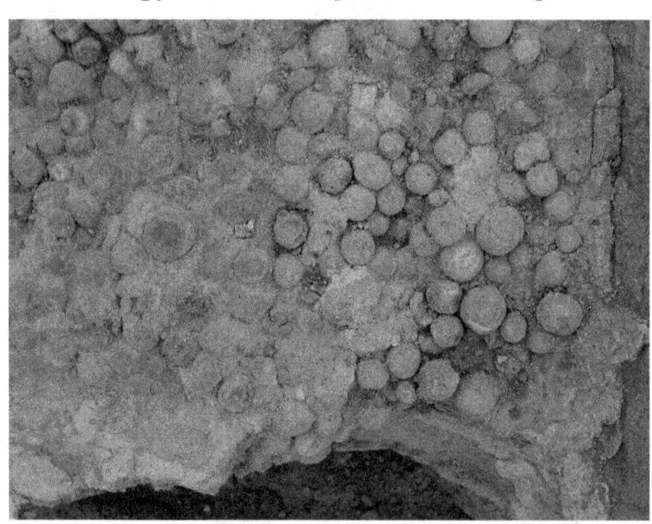

Figure 2.17 A stash of clay balls excavated at Çatalhöyük. [Çatalhöyűk Image Collection File #061401_080517 (1963), shared under a CC-BY-NC 4.0 International License.]

The cook would heat the balls in the house's hearth and then transfer them, probably with stick-tongs, to containers. These were likely clay-lined baskets that held water, bits of meat (usually sheep or goat), and other foods. However, the balls quickly lost their heat in the water and had to be put back on the hearth. Imagine the cook in every family carefully monitoring the movement of several balls back and forth from fire to basket for each cooked meal, making this a tedious and labor-intensive daily task.

The early balls were made of the same fiber-tempered backswamp clays as the mudbricks. And the same paste was used to make the earliest clay vessels, appearing around 7000 BCE. However, this fiber-tempered pottery was unsuitable for cooking, so these early pots probably functioned to serve food or drink. As such they did not directly modify the entanglement with cooking balls.

Nevertheless, the use of clays was changing. Digging for the siltier backswamp clays exposed underlying sandy clays. These clays did not require the addition of organic temper, and they were more efficient at heat-transfer than the fiber-tempered pastes.[24] By about 6600 BCE, the clay balls started to diminish in frequency as larger, thinner, sandier clay pots appeared. These typically show exterior smudging, which indicates they were placed directly on a hearth as cookpots (Figure 2.18).

Cookpot Consequences

Cooking food in a pot frees the cook from having to constantly reheat the clay balls to do other tasks. This change in cooking technology modified the scheduling of labor for domestic activities. It would have transformed gender relations and the division of labor within the household, assuming that women and girls were in charge of food preparation. Ceramic cooking vessels also required more skill, investment of labor, and new resources, including non-local clays and fuel for firing pottery (see Sassaman, "Ceramics").

Figure 2.18 This vessel excavated from a later occupation level at Çatalhöyük shows the marks of having been put over a fire for cooking. [Catalhoyuk Image Collection File #20020801_mal_041 (2002), *shared under a* CC-BY-NC 4.0 *International License.*]

Thus, one form of the entanglement of clay gradually replaced another over several centuries, with profound reverberations for Çatalhöyük society. It was at this same period of transition that the settlement reached its greatest extent and was most densely packed with houses, now made of the larger, sandier mudbricks.[25]

Material Lessons

The entanglement of clay at Neolithic Çatalhöyük illustrates more general insights for understanding the relationships between humans and materials, and the impacts of materials on society. These include how people engage with properties of materials in production processes, the critical difference between potential and actualized properties, and the recognition that some properties are advantageous while others are disadvantageous. In treating materials as bundles of properties, the notion of a "thing" as gathering or assembling is again revealed as a useful way to comprehend the interactions between people and materials.

Materials? Or Properties?

A long-standing bias implicit in Thomsen's Three-Age System, which introduced this chapter, is the notion that stone, bronze, and iron are homogeneous material categories that peoples throughout time and space perceived in an equivalent manner. That is, his scheme narrows our attention to materials as seemingly defined strictly by certain natural, essential qualities. In criticizing this bias, archaeologist Chantal Conneller argues to the contrary that people engage with the *properties* of materials, not with some universally recognized substance in nature (Figure 2.19).[26]

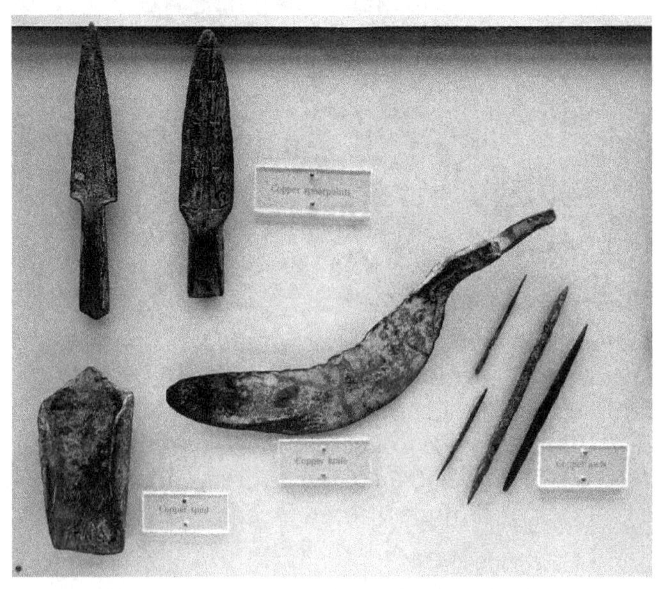

Figure 2.19 Copper knife, spearpoints, awls, and spade, from the "Old Copper Complex" of the western Great Lakes, Late Archaic (pre-farming) period, 3000–1000 BCE, in the Wisconsin Historical Museum. [Photo by user Daderot (2013), shared under a CC0 license. Wikimedia Commons.]

Importantly, specific properties of a material will actually vary depending on human experiences with it and with the other materials and objects brought into relationships with it. These relationships include making comparisons and contrasts between materials—how is clay like or unlike stone (or metal)? They also more literally refer to physically combining materials (e.g., clay and water) or manipulating them with tools (e.g., polishing dried clay objects such as mortar or pots to harden the surface).

Thus Conneller can assert in the above epigraph that "there is no such thing as 'stone'." Instead, a variety of materials may be lumped together at different times and in variable situations by the word "stone." Alternatively, materials we would treat as all "stone"—running the gamut from talc to diamond—might be distinguished as different substances by other peoples.

"Making"

Materials, and the objects made of them, come "bundled" with multiple potential properties. "**Making**"—which includes "unmaking"—is an umbrella term introduced by anthropologist Tim Ingold to encompass the production processes by which people engage with the bundled properties of materials as part of the various projects they undertake.[27] In some cases, properties are known and are drawn into strategic, intentional plans. In other cases, they emerge as unintended consequences of human practices. Thus, what a material *is* in the technical jargon of a modern scientist is not as relevant to those who use it as what it *does* in particular situations (Figure 2.20).[28]

Because situations will vary, what a material *does* is subject to change. This means that any material should be treated as mutable, variable, and dynamic—not inert, fixed, and static. To comprehend the impact of materials on society we must attend to the activation of the properties of materials in human interactions with them, and also to the historical consequences of those interactions, as the Çatalhöyük case study demonstrates so well.

Potential and Actual Properties

In the process of making, some of the **potential properties** of materials are **actualized** out of that practical experience.[29] However, other properties are not, or not immediately so, and remain virtual or latent, unrecognized or unvalued by people. Some properties may emerge as a consequence of physical processes, such as rust or decay, which are dependent on certain environmental circumstances (In Figure 2.19, the copper has oxidized, turning the surface green.). At Çatalhöyük, only the passage of time revealed the continued shrinkage of the smectitic mudbricks well after they were first sundried.

The actualization of potential properties may also result from transformations that reveal hitherto unrealized effects. For example, we consider clay to be useful because it (1) is easily molded into shape, and (2) can be made into hard and durable objects. However, these potential properties of clay emerge only through the application of a certain amount of water in the first case (Figure 2.20), and additional pyrotechnologies in the second.

In many other instances, the potential properties of materials remain latent because they are not relevant to human projects (making). For example, ancient peoples used iron ores such as hematite and ilmenite as a source of red and yellow pigment, or they polished the minerals to make mirrors. All of this happened long before iron-working was invented or introduced. Thus, "material reality is teeming with virtual or potential qualities or properties which never get actualized."[30] Making is what brings out these potentials. To understand a material requires knowing the history of how its various properties emerged as a result of the changing situations of human encounters with it.[31]

Significantly, actualizing and realizing formerly unobserved or unimportant virtual properties is a source of innovation.[32] For example, humans used clay for millennia in many ways other than by firing it at high temperatures to bring about its potential to make hard, durable, waterproof vessels. Once that technology (ceramic pottery) emerged, it changed the courses of human histories. At Çatalhöyük the innovation of clay cooking vessels (Figure 2.18)—brought about by actualizing sandy clay's potential properties—resulted in major social changes.

Figure 2.20 Properties of materials are engaged in their making or using, as here in the case of a potter at work in Jaura, Madhya Pradesh, India. Note the bowl of water kept next to the wheel. [Photo by user Yann (2009), shared under a CC-BY-SA 3.0 Unported License. Wikimedia Commons.]

Affordances and Constraints

Although materials are teeming with properties, not all of them prove advantageous to whatever human projects they are brought into. Going back to the fired clay (ceramic) example, its durability is usually considered a desirable result of that pyrotechnology, but ceramics are also brittle—they break easily.

An analytical concept for differentiating desirable properties from undesirable or unrealized properties comes from the work of ecological psychologist James Gibson.[33] He devised the term **affordance** to refer to the recognized potential properties—for good or ill—for a particular set of actions in a certain situation or environment. Subsequent researchers have modified this term to distinguish advantageous properties—such as the durability of ceramics or the thermal properties of clay cooking balls—from those that create **constraints** on human action. Examples of constraints are the fragility of pottery that mandates careful handling, and the heavy weight of wet mudbricks that increases transport costs.

Affordances as recognized beneficial properties play a disproportionate role in the manufacture or use of objects, but they are always dependent on their context. This means that affordances must be readily apparent to the humans involved in that context or situation. And because humans do not act in isolation but rather in cooperation with others, affordances have a social aspect. Not everyone will agree on whether properties are advantageous or not, so any material or object affordance may require social negotiation.

Backswamp Mudbricks: Affordance or Constraint?

We can only imagine the discussions and disagreements to establish the first houses of what would become the *höyük* now called Çatal. Should the pioneer settlers erect their mudbrick dwellings directly on the deposits of backswamp clay, which would put them at risk of flooding? Or, should they build on the higher natural rises, taking them out of the flood zone but at a greater distance from the backswamp clays preferred for bricks?

We know the answer only in historical retrospect. Note that in this example, affordances include the location and abundance of backswamp clay. They are not limited to some intrinsic properties of clay as scientifically defined, because affordances are situational.

Thus affordances are produced or become evident out of human interactions (making) with materials in a particular context. This means that affordances are dynamic or changeable because humans, materials, and situations will vary in time and space.[34] Archaeologist Chris Doherty argued that the diversity of clays available at Çatalhöyük seems obvious to us today, but that affordance was not realized until the inhabitants began removing the backswamp clay. In so doing, they disrupted the natural environment, changing the affordances of the original landscape.[35] The concept of affordance is another reason to avoid treating materials as stable, universally defined categories.

Things as Assemblages

In similar fashion to Martin Heidegger's dynamic and mutable conception of a "thing," other 20th-century philosophers have referred to materials or objects as "assemblages" or "networks."[36] Individual materials are assemblages of their particular potential properties. For example, at Çatalhöyük backswamp clay, marls, and sandy clays bundle different properties.

The fluid nature of materials, with their individual histories of potential and actualized properties, as well as the constant negotiation of those properties in social projects, impact the objects made from them. Objects are "bundles" of different materials brought together in a certain way, their properties emerging from dynamic and situational human interactions. Although we tend to view objects as solid and stable, they are just as changeable as the materials that compose them. The mudbricks that the residents of Çatalhöyük depended on for 2,000 years exemplify this instability.

Objects as assemblages are unstable in another way. They do not naturally endure in terms of their original networks of associations or meanings, even though some of their physical components may persist. The people of Çatalhöyük regularly razed their houses, transforming them into platforms for subsequent dwellings on top. In some cases these former-houses-as-foundations became places of the dead, with new meanings and values. Thus the mound endures into the present some 9,000 years after it was started; yet, as a network of things, people, and values, it was constantly changing (see Figure 2.11).

The users of objects are generally aware of the inevitable changes they undergo. The Çatalhöyük data reveal how much the people labored, sometimes to an extraordinary degree, to stabilize mutable objects and try to maintain the networks in which the objects participated. These efforts indicate the interdependence of people and materials, and the assembled objects made from them, and how these dependencies played out over thousands of years. Appreciating this complicated history is important to more fully comprehending the impact of materials on society.

Conclusion: Entanglement and its Consequences

The entanglement of clay at Neolithic Çatalhöyük thus provides two important lessons for understanding the impact of materials on societies. The first is that entanglement of materials is a historical process, meaning it has material and social consequences that play out over time. The entanglement begins as materials are deployed to meet human needs and desires. Clay was essential for many daily life-sustaining activities. However, at Çatalhöyük there was no such thing as "clay." The inhabitants differentially utilized the multiple clay-bearing deposits, both in the Çatalhöyük vicinity and from outside the area, according to their particular perceived properties and the contexts of their extraction, transport, processing, and use or reuse.

These clays, and the non-clay materials with which they were naturally or humanly assembled, had their own properties. Some properties were actualized, while potential others remained latent or virtual. Some properties were advantageous to human projects. Besides being abundant and easy to acquire, clay could be formed into multiple, relatively durable objects with a minimum of skill or tools, and could dry in the natural heat of the sun. Other properties posed constraints, such as the weight of wet clay and the high shrinkage rate of the smectitic clay. Both actualized and virtual, advantageous and disadvantageous properties, in their proper contexts, must be thoroughly accounted for in assessing the impact of clay on the Neolithic societies that were ancestral to our own.

This case study further reveals clay objects as assemblages, "things" that participated in assembling higher order things. They gathered different materials, people, processes, and places in their making and unmaking. Things depended on other things and they depended on the people who depended on them.

The second lesson is that entanglement is a historical process whereby human actions are intertwined over time with physical forces. The latter include decay, degradation, corrosion, transformation, wearing out, and running out of materials. When things start to fall apart, the usual reaction is to fix them or find equivalent replacements because entanglement is an entrapment.

This means that all these intertwined processes play a central role in social change. New materials or innovated actualized properties of existing materials are selected for use only if they fit within the existing entanglements. Otherwise they may be ignored. Cook pots at Çatalhöyük had to fit into the existing technology of cooking with clay and making clay vessels, together with the gradual substitution of sandier clays for the siltier backswamp clay.

Through the long lens of archaeology, we can begin to understand the consequences of entanglement at Çatalhöyük. Exactly the same processes are occurring in our lives today, constraining our alternative futures. However, because we are "trapped" by our entanglements, with many new materials now, it is difficult to apprehend how much we are dependent on things that depend on us. Even when we recognize our interdependencies with materials, it is a challenge to overcome them if our entanglements hold us back and prevent us from adopting new materials or alternative technologies for societal needs.

Discussion Questions

1. Select an "earthy" material that is critical to our modern society (e.g., precious and utilitarian metals, fossil fuels, rare earths). Using the four premises of the entanglement model, explain in detail our interdependencies with that material. How can we potentially escape the entrapment of the entanglement of this material?
2. Pick another "earthy" material and explain which of its properties are

"affordances" and which are "constraints." Remember that affordances and constraints are context-dependent and not always inherent in the material itself, so you must specify the context or situation. Were the constraining properties known at the time objects of that material were made or first used? What about potential affordances?

3. Understanding a "thing" as an assembly or gathering is important to comprehending new approaches to how humans use materials and to Hodder's entanglement model. Using Figure 2.15 as an example, diagram the clay cooking ball as a "thing" in the early history of Çatalhöyük. Consider how it assembled other substances, people, places, tools, and processes as it was made and as it was used.
4. Clay at Çatalhöyük was used to make primarily "subceramic" (unfired) objects that were dried in the heat of the sun. The properties of ceramics are generally considered to be superior to those of unfired clay, yet unfired clay does have important advantages or affordances. What were the specific affordances of unfired clay for the people of Çatalhöyük? What purposes continue to be served by unfired clay materials in our modern society?

Key Terms

entanglement
Neolithic
Soil Revolution
clay
Age of Clay
plasticity
temper
thing
paste
tanglegram
making
potential properties

actualized
affordance
constraints

Author Biography

Susan D. Gillespie is Professor of Anthropology at the University of Florida. She received her PhD in Anthropology from the University of Illinois at Urbana-Champaign in 1983, specializing in the archaeology and ethnohistory of Mesoamerica. Dr. Gillespie has directed archaeological projects in the states of Oaxaca and Veracruz, Mexico, investigating the rise of complex society in the Formative Period (ca. 1500–500 BCE). She is the author of *The Aztec Kings: The Construction of Rulership in Mexica History* (Univ. of Arizona Press, 1989), awarded the 1990 Erminie Wheeler-Voegelin Prize by the American Society for Ethnohistory. She co-edited *Things in Motion: Object Itineraries in Anthropological Practice*, with R. A. Joyce (School of Advanced Research Press, 2015), *Archaeology Is Anthropology*, with D. L. Nichols (*Archeological Papers of the American Anthropological Association* No. 13, 2003), and *Beyond Kinship: Social and Material Reproduction in House Societies*, with R. A. Joyce (Univ. of Pennsylvania Press, 2000).

Notes

1. Chantal Conneller, *An Archaeology of Materials: Substantial Transformation in Early Prehistoric Europe* (New York: Routledge, 2011), 82, http://www.worldcat.org/oclc/1040885381.
2. On Thomsen and the impact of his and subsequent work in the periodization of early history in Europe, see Bruce G. Trigger, *A History of Archaeological Thought*, 2nd ed. (Cambridge: Cambridge Univ. Press, 1989; 2006), 121–29, 147–48, http://www.worldcat.org/oclc/1102556484.
3. Suzanne Staubach, *Clay: The History and Evolution of Humankind's Relationship with Earth's Most Primal Element* (New York: Berkley Books, 2005; Hanover, NH: Univ. Press of New England, 2013), http://www.worldcat.org/oclc/864745156. Refers to UPNE edition.
4. The model of entanglement used in this chapter was the invention of archaeologist Ian Hodder as explained in several of his publications: "Human-thing Entanglement: Towards an Integrated Archaeological Perspective," *Journal of the Royal Anthropological Institute* (N.S.) 17 (2011): 154–77, https://www.jstor.org/stable/23011576.; *Entangled: An Archaeology of the Relationship between Humans and Things* (Malden, MA: Wiley-Blackwell, 2012), http://www.worldcat.org/oclc/1151155585.; and "Becoming Entangled in Things," in *Substantive Technologies at Çatalhöyük: Reports from the 2000–2008 Seasons*, ed. Ian Hodder, Çatalhöyük Research Project Series Vol. 9 (London: British Institute at Ankara and Los Angeles: Cotsen Institute of Archaeology Press, 2013), 1–25, http://www.worldcat.org/oclc/986819635.
5. In his 1934 book, *New Light on the Most Ancient East*, renowned archaeologist Vere Gordon Childe proposed that humanity experienced two "revolutions": the Neolithic food-producing

revolution followed by the Urban Revolution that coincided with the Bronze Age in the Old World (in Trigger, *History of Archaeological Thought*, 324).

6. Julian Thomas, "An Economy of Substances in Earlier Neolithic Britain," in *Material Symbols: Culture and Economy in Prehistory*, ed. John E. Robb, Occasional Paper No. 26 (Carbondale, IL: Center for Archaeological Investigations, Southern Illinois Univ., 1999), 75–76, http://www.worldcat.org/oclc/41561725.
7. Nicole Boivin, "Geoarchaeology and the Goddess Laksmi: Rajasthani Insights into Geoarchaeological Methods and Prehistoric Soil Use," in *Soils, Stones and Symbols: Cultural Perceptions of the Mineral World*, eds. Nicole Boivin and Mary Ann Owoc (London: Univ. College London Press, 2004), 174, 181, http://www.worldcat.org/oclc/56911728.
8. Mirjana Stevanović, "The Age of Clay: The Social Dynamics of House Destruction," *Journal of Anthropological Archaeology* 16 (1997): 334–95, https://doi.org/10.1006/jaar.1997.0310.
9. Chris Doherty, "Sourcing Çatalhöyük's Clays," in Hodder, ed., *Substantive Technologies at Çatalhöyük*, 51.
10. Boivin, ""Geoarchaeology and the Goddess Laksmi," 177.
11. Boivin, 178.
12. James Mellaart, *Çatal Hüyük: A Neolithic Town in Anatolia* (New York: McGraw Hill, 1967), http://www.worldcat.org/oclc/582727604. See also Ian Hodder, *The Leopard's Tale: Revealing the Mysteries of Çatalhöyük* (London: Thames & Hudson, 2006), http://www.worldcat.org/oclc/1045442166; Hodder, *Entanglement*, 60.
13. Serena Love, "An Archaeology of Mudbrick Houses from Çatalhöyük," in Hodder, ed., *Substantive Technologies at Çatalhöyük*, 96; Burcu Tung, "Building with Mud: An Analysis of Architectural Materials at Çatalhöyük," in Hodder, ed., *Substantive Technologies at Çatalhöyük*, 78.
14. This section draws on the analysis of Çatalhöyük's clay sources and how they were used by its inhabitants, conducted by archaeologist Chris Doherty; see Doherty, "Sourcing Çatalhöyük's Clays."
15. The premises of Hodder's model are best explained in Hodder, "Human-Thing Entanglement," and Hodder, *Entangled*.
16. Martin Heidegger (1889–1976) was a leader in the 20th-century philosophy of Continental Phenomenology. His influential essay, "The Thing" (*Das Ding*) was published in English in Heidegger, *Poetry, Language, Thought*, trans. A. Hofstadter (London: Harper, 1971), 165–82, http://www.worldcat.org/oclc/1070903602.
17. For tempering of the mudbricks, see Hodder, *Entangled*, 152; Mirjana Stevanović, "New Discoveries in House Construction at Çatalhöyük," in Hodder, ed., *Substantive Technologies at Çatalhöyük*, 111; and Tung, "Building with Mud," in Hodder, ed., *Substantive Technologies at Çatalhöyük*, 79.
18. Hodder, "Human-Thing Entanglement," 157.
19. Doherty, "Sourcing Çatalhöyük's Clays," 64; Hodder, "Human-Thing Entanglement," 160–61; Hodder, *Entangled*, 65–67.
20. Hodder, *Entangled*, 66.
21. Hodder, *Entangled*, 67.
22. Hodder, *Entangled*, 65.
23. Hodder, *Entangled*, 152–56.
24. Doherty, "Sourcing Çatalhöyük's Clays," 65.

25. Hodder, *Entangled*, 153–54.
26. Conneller, *An Archaeology of Materials*, 5–8.
27. Tim Ingold, *Making: Anthropology, Archaeology, Art and Architecture* (London: Routledge, 2013), 20–21, http://www.worldcat.org/oclc/840416927.
28. Conneller, *An Archaeology of Materials*, 19.
29. This is the distinction 20th-century philosophers have made between potential and actual (Alfred North Whitehead, 1861–1947), or virtual and actual (Gilles Deleuze, 1925–1995) properties. See Gavin Lucas, *Understanding the Archaeological Record* (Cambridge: Cambridge Univ. Press, 2012), 167, http://www.worldcat.org/oclc/1162097299.
30. Lucas, *Understanding the Archaeological Record*, 167.
31. Tim Ingold, "Toward an Ecology of Materials." *Annual Review of Anthropology* 41 (2012): 434–35.
32. Lucas, *Understanding the Archaeological Record*, 167.
33. James J. Gibson (1904–1949) was a leader in the psychology of visual perception. See Gibson, *The Ecological Approach to Visual Perception* (Boston: Houghton Mifflin, 1979), http://www.worldcat.org/oclc/1000427293. For revisions of his ideas see Carl Knappett, "The Affordances of Things: a Post-Gibsonian Perspective on the Relationality of Mind and Matter," in *Rethinking Materiality: The Engagement of Mind with the Material World*, eds. Elizabeth DeMarrais, Chris Gosden, and Colin Renfrew (Cambridge: McDonald Institute for Archaeological Research, Univ. of Cambridge, 2004), 43–51, http://www.worldcat.org/oclc/60740603. See also Hodder, *Entangled*, 48–50.
34. Knappett, 46; see also Conneller, *An Archaeology of Materials*, 1; Ingold, "Toward an Ecology of Materials," 435.
35. Doherty, "Sourcing Çatalhöyük's Clays," 65–66.
36. Lucas, *Understanding the Archaeological Record*, 188; Ingold, "Toward an Ecology of Materials"; Bruno Latour, *Reassembling the Social: An Introduction to Actor-Network Theory* (Oxford: Oxford Univ. Press, 2005), http://www.worldcat.org/oclc/1156912903.

Ceramics: Firing Clay and Flaking Stone

KENNETH E. SASSAMAN

> "Failure with clay was more complete and more spectacular than with other forms of art. . . Any one of the old four—earth, air, fire, water—can betray you." —A.S. Byatt, *The Children's Book*

Abstract

This chapter follows from the previous chapter on earthy materials to consider the social impacts when transforming clay and rock into glass and other **ceramic** substances. Introduced is the concept of **operational sequence**, or the process by which affordances of both materials and societies are assembled and disassembled as things are made, used, and discarded. In particular, this chapter contrasts how glass-like rock, such as obsidian and flint, can be broken with human force with the firing of clay to produce true ceramics. The operational sequence for making both an Ice-Age spear point and a ceramic pot illustrates the contingent relationships of physical and social acts in making things, while also showcasing the evolutionary conditions under which ancestral humans developed the cognitive, motor, and social skills to achieve particular outcomes from an array of possibilities. The application of thermal energy to first stone and then clay introduced additional affordances, as well as constraints, that inform our understanding of the potential for ceramic materials of the future. Examining the relationship of human energy application to thermal energy application in different ceramics also provides important lessons for how we can use future ceramics to store and generate energy at lower costs and with fewer negative impacts than conventional technologies.

Introduction

Ceramics are among the premier materials that connect the distant past of our earliest *Homo sapiens* ancestors to our future. From the first pottery vessels of ancient China to the bone implants and fuel cells of tomorrow, ceramics have been developed to solve human problems for over 20,000 years. With such a long history of research and development, ceramics embody many impacts of materials on society. Since their beginning, ceramic materials served social needs, such as preparing a meal to share with others on a plate or decorating a place of ritual gathering with a ceramic mask. Future applications in communications, medicine, and energy production ensure that ceramics will remain integral to human societies for generations to come. As has long been the case, innovations in ceramics arise from the novel interplay of the properties of different substances—clay, temper, water, flux—and the application of heat. But they also arise from changing relationships between producers and consumers, experts and novices, and men and women. Before we delve into these sorts of issues we might first ask, what exactly is a ceramic?

What Makes Something a Ceramic Material?

That plate in your kitchen sink and tiles on your bathroom floor are most likely ceramic. So too are the insulators of your light bulbs, components of your microelectronics, the brake linings of your car, and maybe even the crown on your tooth. The wide array of applications for ceramics in our past and those of the future make it difficult to define ceramics in simple terms. We can agree that ceramics are inorganic solids composed of a combination of metals and non-metal or metalloid atoms that result in nonmetallic materials. We might add that in many cases they are a refractory, that is, a substance resistant to heat. It takes a great deal of heat to make a true ceramic, and some ancient ceramic vessels were designed to convey heat efficiently, as with the wet cooking of foods like corn and barley that required prolonged simmering to make them edible.

To sort through the array of materials broadly classified as ceramic, it is useful to distinguish between traditional and technical (or functional) forms of ceramics. Traditional ceramics include objects made from clay. A by-product of the weathering of rock, clay is a fine-grained material consisting of often alumina silicates with traces of metal oxides and organic matter. To reiterate key properties of clay (see Gillespie, "Clay"): (1) size is critical (particles must be less than two microns in size); (2) the water content of clay makes it plastic, capable of being molded while still wet; (3) dried and fired clay becomes hard and brittle; and (4) most clays shrink when dried, causing cracks to form at surfaces and requiring temper to prevent cracking. The utility of clay for humans depends on how its properties are recognized and manipulated, including combining it with other substances (e.g., water, temper), and usually subjecting it to heat.

Traditional ceramics are typically based on the firing of clay. A clay can become a ceramic only if subjected to temperatures in excess of about 1,200 degrees Celsius. The outcome is known as **vitrification**, or the fusing of the layers of clay together with a glass. The first potters to achieve this goal lived in ancient China, under the Han Dynasty (ca. 200 BCE–220 CE); their early porcelain was not only glassy in composition but also translucent because they used white kaolin clay. Pottery making for millennia before and since the dawn of porcelain involved the production of terra cottas, earthenwares, and other forms of **subceramics**, meaning they were not vitrified in the firing process because the temperature was not hot enough to permanently bond the clay layers. For example, clay could simply be sunbaked to achieve a relatively durable form for purposes such as hot-rock cooking or house construction.

Technical forms of ceramic (functional ceramics) go well beyond the vitrification of clay to include carbides, pure oxides, and nitrides, among other materials. Ceramics can be either **crystalline** or **amorphous**, whereas glasses by definition are always amorphous (lacking long-range crystal structure). Technical ceramics can utilize their crystal structure and composition to create new, interesting properties and can also include glasses other than silicates (e.g., chalcogenides, tellurites, gallates, germanates, heavy metal oxide glass). A naturally occurring silicate glass known as obsidian that was used for millennia as raw material for stone tools is not usually considered a technical ceramic. In fact, while materials engineers consider glasses a subset of ceramics, archaeologists generally do not lump glass and ceramics together, as each has a distinct cultural history.

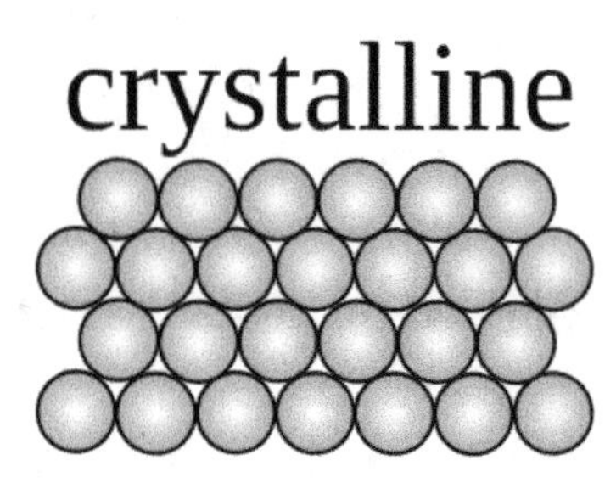

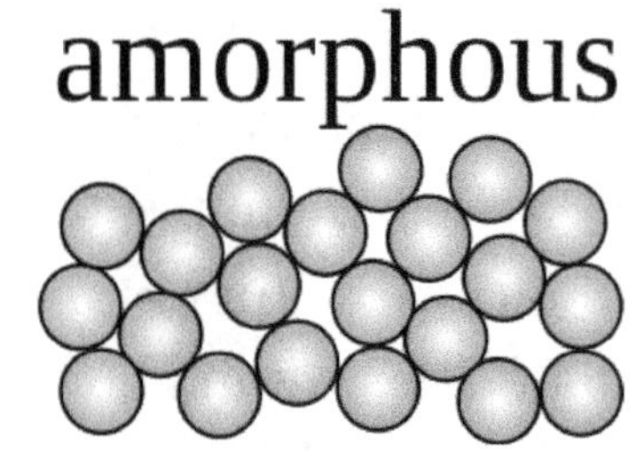

Figure 3.1 The difference at the atomic scale between a crystalline and an amorphous solid. By definition a crystalline solid exhibits long-range order, but an amorphous solid does not. [Image by user Cdang (2011), shared under a CC-BY-SA 3.0 Unported License. Wikimedia Commons.]

For the purpose of this chapter, we will focus not on the physical or chemical composition of ceramics but instead on their production and use. We are most concerned here with the transformation of matter in the manipulation of ceramics, a process that entails the harnessing of energy, which has also entailed hidden costs and unintended consequences for societies of all sorts.

Operational Sequences

A useful way to begin discussion of the impact of ceramic materials on society is to examine the production process. Here we mean production processes in general, not any particular one. All such material production processes have operational sequences, which are a series of decisions and actions that lead to intended goals. We of course follow operational sequences in many of the things we do, from cooking food, to painting a room, to repairing a bicycle tire. And we all know, from experience, what happens when a sequence is enacted out of order, especially when processes have irreversible outcomes. For example, if you add the eggs to a brownie after it's cooked, it's probably not palatable.

From a humanistic perspective, there is much more to operational sequences than the step-by-step procedure for getting something done. Doing work and making things involve the interplay between persons and matter, notably the bodily and social acts involved in production. Anthropologists often employ the concept of ***chaîne opératoire*** (French for operational sequence) to describe how the technical, bodily, and social acts of production are mutually dependent and collectively the basis for innovation and change. In this sense, we might say that society is made as things are made, or that production involves entanglements that reach deep into the fabric of society and its cultures (see Gillespie, "Clay").

In their study of artifacts, archaeologists develop models of operational sequences to characterize the production of stone and bone tools, pottery, and rock art, among other things. This aids them in understanding how the sociological aspects of the objects. According to archaeologist Peter Bleed, such models take two general forms.[1] Some simply describe the sequence of operations towards a predetermined goal, much like recipes in a cookbook. Others emphasize the potential for variation in sequences by looking at how technical and social issues interact dynamically. This latter approach reveals how change occurs over time and thus provides some basis for imagining how future innovations in ceramics and glass may further impact societies. Let's take a look at the possible futures of glass by first looking back to the ancient past.

Breaking Ancestral Glass

Although our ancestors manipulated any variety of organic and inorganic substances since the beginning of human time, things made from stone comprise the oldest archaeological evidence for making things. Stones used for early tools were more durable than other substances humans modified (like wood) so by default stone tools are often the only materials we have to investigate the beginnings of human engineering. We have in the archaeological record of Africa, for instance, evidence for the making and using of flaked stone tools going back over three million years, well before the appearance of fully modern humans at about 200,000 years ago.

To call a stone tool a **flaked** stone tool is to indicate it was made by the removal of flakes from a parent core of rock. Flakes can be removed from cores with either percussion (knocking them off) or by applying steady pressure (pushing them off). How the stone breaks depends on a number of things, most notably what geologists call **cleavage planes**: the planes of relative weakness in crystalline structures. Rocks differ in their type of cleavage. Halite and galena break into cubes, calcite breaks into rhomboids, others into prisms, and so on. All of these geometric fracture tendencies tend to limit the shapes one can achieve through fracture. To achieve more complex shapes, rock that is lacking in any predetermined breakage pattern is needed.

Worldwide, rocks that lend themselves to "unnatural" breakage consist of microcrystalline quartz (silica), such as chert, flint, jasper, agate, and chalcedony, as well as a variety of lesser materials like quartzite and rhyolite.[2]. In a class unto itself is obsidian, a glass-like rock that forms when felsic lava (feldspar and quartz) is cooled rapidly after being spewed from a volcano. Rapid cooling minimizes crystal growth, resulting in an amorphous glass. Obsidian like the microcrystalline materials mentioned above is very isotropic, which enables complex shape fabrication as will be discussed in the next section. With the right application of force or pressure, obsidian can be flaked to produce edges sharper than the sharpest metallic surgical scalpels.[3] In fact, obsidian surgical blades are commercially available.

Sequencing the Production of a Clovis Point

So let's pick a traditional shape of stone tool and run through the technical operational sequence for its production. Our shape of choice is a Clovis point (Figure 3.2), a lance-shaped blade that was flaked on both sides (making it a biface) to achieve a thin, lenticular cross-section.

Next a groove or **flute** on the bottom of the piece needed to be fabricated in order to enable its attachment to a handle or shaft. Clovis points were made in North America from about 13,200–12,900 years ago by hunters of mammoth, mastodon, and other Ice Age creatures.[4] Modern **flintknappers** (persons who flake, or knap, stone) have replicated Clovis technology and attest to the high level of skill involved, particularly in removing the distinctive flutes on either face at the base, a final, risky step in the process that is successfully executed only if all other steps are followed in proper sequence.

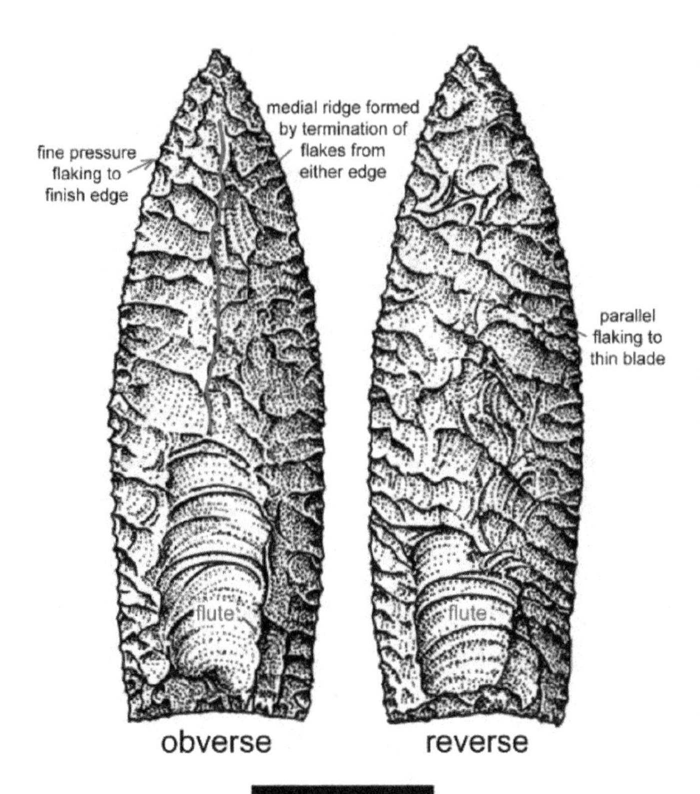

Figure 3.2 Drawing of two sides of a replicated Clovis point, showing some of the diagnostic features mentioned in the text. [Adapted from illustration by Melanie Diedrich of replica made by Scott Williams. Internet Archive.]

Acquire Raw Material

It goes without saying that the first step in the production of Clovis point must be the acquisition of raw material, in this case stone. But as discussed above and in the sidebar below, not any old stone will do. The technical requirements for Clovis points demand the highest-quality raw materials. North America is blessed with all sorts of quality stones for tool making (toolstones) including obsidian, but geological sources are scattered widely across the continent and some locations that were excellent for hunting and dwelling were devoid of stone. Thus, for Clovis tool makers, getting rock meant traveling to sources, either directly from places of dwelling or in the course of moving from place to place over the year. Many of the points made by Clovis hunters were displaced from the geological sources of their raw materials by hundreds of kilometers.

Activity: Watch a video

Jeff Boudreau, Massachusetts Archaeological Society, demonstrates how to make a Clovis Point in this PBS video (https://www.youtube.com/watch?v=jRax_a8t4C4&t=7s).

- Which tools were typically used to fabricate a Clovis point?
- Was it necessary to sharpen the Clovis point?

- What was the most difficult part of fabricating a Clovis point?

Prepare Raw Material for Transport

If you have to travel far from home to get rock, you probably do not want to carry back useless material. Archaeologist Charlotte Beck and colleagues researched the costs of transporting rock from quarries to places of habitation and found that ancient toolmakers minimized their transportation costs by reducing raw nodules of rock into blanks suited to further reduction.[5] The greater the distance between source and home, they found, the greater the effort to remove rocky mass of limited value. Minimally, this involved removing the cortex or **exterior layer** of the rock, which is not terribly conducive to flaking, as well as protuberances and other irregularities.

How Can You Tell an Actual Stone Tool from Some Old Rock?

The late-20th-century comic strip *Calvin and Hobbes* occasionally featured the efforts of its human protagonist to make big discoveries by digging into the earth. The few dirty rocks Calvin found in the strip above were treasure to him, but were they artifacts—that is, objects of human modification? In the case of flaked stone tools, the evidence for human agency is distinctive. Fine-grained rocks that are isotropic or amorphous are usually shaped by removing flakes from the surfaces that converge at edges. If you apply a force to an edge you drive off a flake from one of more surfaces that is elongated, and potentially very thin (instead of radial in shape if the force is applied naturally toward the core). Experts in flake-stone tool manufacture (flintknappers) can manipulate the shape of the edge and angle of applied force to essentially "sculpt" the surface of a core, one flake at a time. Archaeologists are trained to recognize the attributes of controlled reduction and can therefore distinguish manufactured tools from "a few dirty rocks," as well as the by-products of manufacture, notably the many flakes that are removed in sequence to achieve a desired end. Sure, rocks occasionally break in ways that mimic human agency—for instance, when cobbles of chert or obsidian impact other stones during a landslide. But the more steps involved in the reduction of a core, the lower the chance a particular fracture pattern could be mimicked by natural agents. For these and many more reasons, the "treasure" archaeologists find in stone tools is the capacity they have for revealing so much about human societies of the ancient past.

Prepare Core for Reduction

Once you acquire rock of acceptable quality and transport it home, it is time to shape it into a core—a somewhat cylindrical starting piece—of appropriate form and size. At this stage of reduction, the primary objective is to preform the final product, or simply "rough it out." It will take a hammerstone or billet (i.e., an antler or bone hammer) to achieve this objective, along with good planning and motor skills. If the nodule of rock is large enough, a large flake can be struck from its surface and used as a preform or starting point for your Clovis point. Otherwise, the preform resides inside the core itself, and can be revealed only through the careful removal of multiple flakes from across the surface. This is hardly a random or haphazard process, but instead one whose successive steps are contingent upon previous steps.

Shape and Thin Biface

The contingencies of core reduction become even more critical as you approach the final shape of the intended product. Like many flaked stone knives and projectiles, Clovis points are thin in cross-section; they have to be not only because an acute edge is needed to cut and pierce animal tissue, but also because a thin cross-section is conducive to edge maintenance, resharpening the edge to enable its ongoing use and to increase the longevity of the tool. Clovis flintknappers often used an "overshot" technique to drive flakes across the entire face of a core, a decidedly tricky move. Ultimately, however, they had to leave an elevated ridge along the midline of each face of the tool in order to accommodate the bottom flutes yet to come. To do this they had to terminate flakes at the midline, halfway across the surface, an outcome best achieved by applying pressure to the edge of the tool with an antler tip or some such implement to remove thin, narrow flakes.

Remove Flutes

Now comes the fateful moment, when the distinctive flutes of Clovis points are removed from the base of the tool on either side. Fluting will only be possible if the bifacial core has been prepared to precise tolerances. It is risky business indeed, and few modern flintknappers have mastered the technique. Experiments in percussion, pressure, and even the use of levers show that flutes can be removed in a variety of ways. Most impressive perhaps is when a flintknapper holds the biface in his or her palm and drives off a flute with the free swing of a hammer.

Finalize Edges

Now the edges have to be finalized by removing small flakes along the margins from either side. This too is done with pressure, or what is known as "pressure flaking." Holding the tool in the palm of one's hand, an antler tine or tip is applied to the edge to remove flakes in succession, from base to tip, or tip to base. The result is an incredibly sharp edge, serrated if the tool maker desires. This same technique will be used to resharpen the tool as its edges become dull through use.

Grind Basal Margins

Before we can attach our finished product to a handle or shaft, the basal or bottom edges of the Clovis point must be ground. This will prevent the edges from tearing into the materials used to bind the tool to its handle. An abrasive stone is good for this task, perhaps the same one used to prepare edges for flaking.

Affix to Handle or Shaft

Your finished product is not terribly useful without a handle (if it's a knife), or a shaft (if it's a projectile). Now that you have the basal margins or bottom ground, you can fit it to a shaft of some sort, and this step requires even more materials and know-how. A split wooden handle will accept the fluted surface of your tool nicely, but there are other options available, some involving multiple parts. No matter the option chosen, you will need some sort of binding material, such as sinew from a game animal, and perhaps some type of glue. Blood works well, as does pine resin. Keep in mind that you may want to remove your Clovis point when it breaks or is spent, because that handle was likely time-consuming and costly to make. Tool maintenance and replacement are foremost concerns.

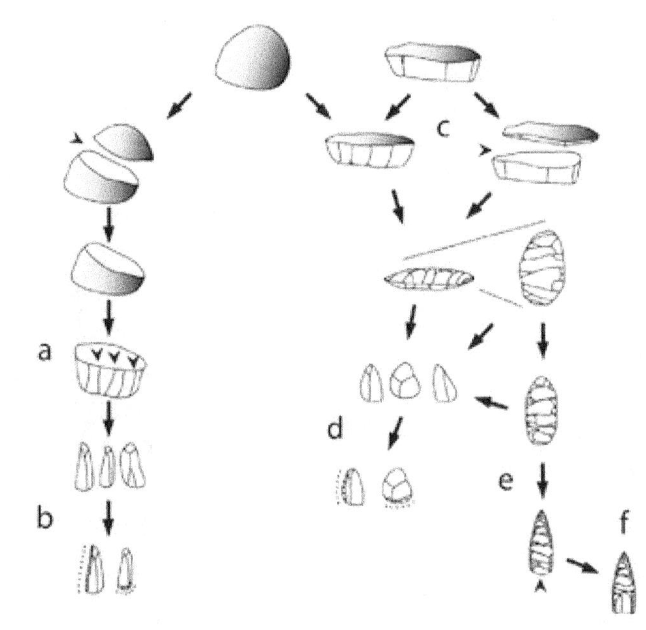

Figure 3.3 Paleoindian Reduction Sequence for fabrication of both flake tools and Clovis Points. [Adapted from Loren G. Davis, Samuel C. Willis, and Shane J. MacFarlan, "Lithic Technology, Cultural Transmission, and the Nature of the Far Western Paleoarchaic/Paleoindian Co-Tradition," in Meetings at the Margins: Prehistoric Cultural Interactions in the Intermountain West, ed. David Rhode (Salt Lake City: U of Utah P, 2012), 47-64, https://muse.jhu.edu/book/41469.]

Social Implications?

We have followed an operational sequence for making a Clovis point, but so what? Could not a single individual follow this sequence alone, from beginning to end, and have a Clovis point to show for their effort? Sure, theoretically. But when we consider that Clovis points were used to dispatch mammoth and other megamammals, we understand that the *application* of these tools was inherently social: mammoth hunting was communal, a social affair, involving many persons, as was the processing and sharing of hundreds of pounds of meat, fat, bone marrow, hide, and other useful products from the animal.

Arguably, each step of the operational sequence for making Clovis points involved social acts. Consider that in acquiring raw material from locations hundreds of kilometers away from the manufacturing, toolmakers crossed over land occupied by others. Alternatively, rock could have been acquired in the course of regional settlement moves, what archaeologist Lewis Binford called **embedded procurement.**[6] In that case, acquisition was embedded in the movements of entire groups, not just individuals. Some archaeologists think that people moved entire settlements or developed territories just to be close to high quality rock.

As we move on to the planned reduction of raw material into cores, the social acts involved cross-generations of toolmakers in networks of learning. Flintknapping is a nuanced skill, one that is not readily assimilated without apprenticeship and mentoring. So too is knowledge of the locations of raw materials, what anthropologists call **landscape learning**. The operational sequence of flaking and the final form of the Clovis point were matters of longstanding tradition, taught and maintained from generation to generation, Innovations arose that led to regional variations in how fluted points were made and used, and over time, as Clovis disappeared as a tradition and was replaced by descendent traditions, ancient knowledge was lost to change. The upshot is that technical know-how in cultures without Google and other literary forms of information sharing was transmitted socially, from expert to novice. The process of learning was situated in the relationships people had to one another; change those relationships and you impact the content and process of learning and vice versa.

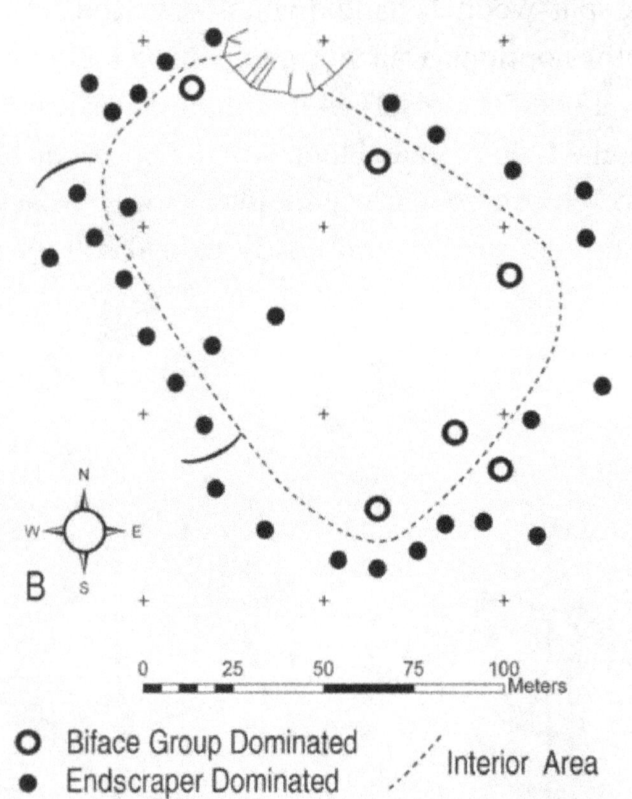

Figure 3.4 Bull Brook Site North of Boston, MA. Clusters of fluted bifaces, end scrapers, and other tools in oval array across area the size of two football fields; Brian Robinson interprets this as a highly structured aggregation settlement associated with fall and winter caribou hunt. [From Robinson et al., "Paleoindian Aggregation."]

And finally, in the application of fluted points for communal hunting and food sharing, core principles of society are revealed. This is exemplified best at an archaeological site in northeastern Massachusetts, at a place called Bull Brook.

From the distribution of lithic (stone) artifacts from this site, archaeologist Brian Robinson has reconstructed in detail a gathering of several dozen Paleoindian hunters and their families occupying this area 10,000–11,000 years ago.[7] Family campsites were arrayed in the large circle. The raw materials for making fluted points and a variety of other stone tools came from various places across the region, signaling a gathering of communities that otherwise spent time apart from one another. They came together to hunt caribou, a migratory herd species that was targeted for communal hunts well after mammoth and mastodons went extinct. To prepare for the hunt, toolmakers crafted many fluted points. Evidently, the risky step of fluting was shared among toolmakers. They gathered in the center of the camp circle to remove flutes, presumably under the guidance of a ritual specialist, someone who knew how to minimize risk and enhance success through many years of experience. Around the perimeter of the camp circle were clusters of flaked-stone scrapers, tools used to process hides for clothing, bedding, and perhaps tent covers. Is the spatial distinction between locations of fluting and hide working indicative of a division of labor? Likely so, and most likely a division along lines of gender, given analogs with historic-era bison hunters of the Plains (i.e., men hunted bison; women processed hides). Attributing particular tasks to particular genders has its pitfalls in archaeology, but such distinctions are nonetheless relevant to our understanding of the implications of materials on society (see Bryant, "Plastics").

Mind, Body, and Society in Evolutionary Terms

Our rather lengthy excursion into the operational sequence of a Clovis point can be put into long-term evolutionary perspective to understand how modern humans came to be different from other species, and how human societies were impacted by transforming matter into useful products. On the first count, the mental and physical ability to make a Clovis point was underpinned by a three-million-year evolutionary history of human ancestry. If you know the classic film 2001: A *Space Odyssey*, you will recall the initial attempts of protohumans or advanced primates to smash bones with hammers (Figure 3.5). Lacking a tradition for tool making and thus without knowledge of an operational sequence for achieving a particular form, their results were haphazard. Still, the connection between cause (application of force) and effect (breaking bone) was apparent to these early tool makers, and, with time—and lots of trial and error—our ancestors came to understand the process of tool and weapons manufacture.

Figure 3.5 An ancestral human in the 1968 film 2001: A Space Odyssey comes to realize through experimentation that the application of force through the instrument of a hammer results in bone breakage.

Captured in this gem of cinematic art is the process by which mind and body evolved in sync to produce a creature capable of not only applying energy (controlled force) to modify matter, but to also anticipate each step of an operational sequence to achieve desired outcomes. The cognitive and somatic developments behind this evolutionary history are beyond the scope of this chapter, but suffice to say that they arrived out of the interplay among planning, motor control, problem solving, knowledge transfer, and memory. By the time we get to the Clovis era millions of year later (only 10,000–13,000 years ago)—well beyond the evolution of fully modern humans—the level of strategic planning is impressive, albeit hidden away in the design and use of fluted points, among other tools. In this regard two aspects bear mentioning: (1) each step of the operational sequence was contingent on the prior step, and the sequence was irreversible, so the pressure was on to get it right; and (2) the costs of making a fluted point were so high that an adequate return on its investment entailed long-term maintenance and even incentives to recycle broken and worn tools into other products, a common practice at locations far from quarries.

Tools were made by transforming matter into usable material forms. But societies were also made in this process of transforming matter. Learning, sharing, cooperating and competing, territorialism, divisions of labor, gender roles, and the flows of goods, services, and personnel are all entangled in the operational sequence of making and using Clovis points. If a process can be this entangled 13,000 years ago among relatively simple, small-scale societies, imagine how much more entangled they are in complex, global-scale societies of the modern era. Conversely, perhaps we are not all that much different than these ancient toolmakers, as the history of making and using pottery suggests.

Harnessing Energy through Ceramics

As discussed earlier, a traditional ceramic is vitrified clay, essentially clay that has been transformed into glass by heat of at least 1,200 degrees Celsius. It takes a **kiln** or furnace to maintain temperatures this high; temperatures in an open-pit fire can exceed this threshold, but air circulation is such that average sustained temperatures rarely exceed 1,000 degrees Celsius (Figure 3.6).

Figure 3.6 An example of open pit firing on a large mound in Kalabougou, Mali (2012). [Photo by Marco Bellucci, shared under a CC-BY 2.0 Generic License. Wikimedia Commons.]

Potters using open-pit firing could **sinter** clay into a hard, relatively durable substance without vitrifying it, with outcomes that we classify today as **subceramics**. Kilns of various designs show up in several places across the globe as early as 10,000 years ago, but designs capable of vitrifying clay date to only the last 2,000 years, the earliest in China, Japan, and the Roman world.

The long history of research and development from subceramics to ceramics is filled with twists and turns as potters "discovered" the latent affordances of clays, tempers, and other substances, playing off and instigating, in many cases, changes in society that inflected the demand for innovations.

Figure 3.7 A beehive kiln in Chewelah, Washington. [Photo by Catherine dee Auvil (1999), shared under a CC0 license. Wikimedia Commons.]

Pottery and Society at the Dawn of Agriculture

The entanglements of clay and society at Çatahöyük (see Gillespie, Clay) were experienced under various circumstances by societies worldwide. Clay was particularly impactful on society when its processing accompanied the advent of agriculture. In places where domesticated grain became the staple of agricultural economies, pottery was needed to realize its nutritional value, to render wild forms of wheat, barley, and rice palatable. Most often this entailed the process of prolonged boiling or simmering. Many such grains require forty minutes or more of sustained boiling to absorb moisture and make them digestible. Pottery conducive to sustained boiling is tricky to make because one has to balance the need for thermal conductivity against the risk of thermal shock, plus pottery is generally a refractory, an insulator, not a great conductor of heat.

As we learned in Gillespie's chapter on clay, before there was pottery at Çatahöyük there were clay balls. Similar technology was used in the American Southeast about 5,000 years ago, before clay pots were invented, and even afterwards for several centuries. The first pottery in the Southeast was designed for hot rock (and clay ball) cooking, and was thus intentionally thick-walled, to insulate internal heat. Similar technology appeared in the American Midwest a millennium later, where local communities began to consume in earnest the wild versions of weedy plants with starchy seeds, one of which (*Chenopodium*) is related to the *quinoa* that is gaining popularity today as a gluten-free grain. Thick-walled subceramics were just fine for traditional, hot-rock cooking, but if wild grains were to find a foothold in the economy, and be set on the pathway to domestication, pots that could be set directly over a fire were needed. That meant overcoming the poor thermal conductivity of clay pots.

Archaeologist David Braun documented the steps Midwestern potters took to unleash the potential of early pottery for prolonged boiling.[8] New operational sequences arose, all bent towards improving the thermal conductivity of pottery while reducing the risk of thermal shock cracking from large changes in temperature. Local clays were sufficient, but increasingly added to clay was fine quartz sand, a substance not only with decent thermal conductivity, but also with a coefficient of thermal expansion slightly greater than most clays, which left, after firing, microscopic voids in the fabric of the pottery walls that arrested cracks before they propagated. Thinner walls also challenged traditional forming techniques, which included simply molding clay into a vessel by hand, or assembling slabs into a vessel shape. A **coiling** technique proved effective, where walls were assembled gradually from bottom to top, like the courses of brick in a building. The walls were then compressed by paddling. The paste had to be neither too wet nor too dry to make this work. Likewise, traditional forms with angular bases and corners would no longer cut it. Jars and pots with angular bases or shoulders gave way to globular vessels; their lack of angles reduced thermal shock, and their tapered openings minimized evaporative heat loss (Figure 3.8).

Social changes attending the rise of agriculture are legion. Communities became less mobile, tethered now to patches of land they modified and to the plants they cultivated. Populations grew as both the demand for labor rose and constraints on fertility waned due to a more stationary lifestyle. Social concepts like property and inheritance emerged to foster multigenerational connections among persons, things, and land. Perceptions of time were altered to accommodate the delayed return on investments that come from farming and food storage. Greater divisions of labor appeared to meet the increasingly specialized demands of production and distribution.

Intensify!

Figure 3.8 An ideal cookpot, in this case from the Intermediate Bronze Age of the southern Levant of the Middle East. [Image by Foundation for Archaeological Research of the Land of Israel].

Above all, conditions were in place for what anthropologists call **intensification**, which means simply to increase production, often at increased costs per unit. Pottery was at the forefront of intensification, with numerous innovations addressing the growing demands of larger and more sedentary populations. Yes, communities had more mouths to feed. They also had other social incentives driving increased food production such as public works, military capacity, and institutions of government and religion. Eventually, with the rise of markets and commerce, the production of pottery, like so many other commodities, became specialized. Operational sequences not only became more complex—with the addition of more steps and more substances, like glazes and fluxes—but also segmented and distributed among different people, places, and schedules. Gender roles and relations were most directly affected in some cases, as in the **commodification** of pottery outlined below.

Key Concepts: Changing Gender Relations and the Commodification of Pottery

Visit Acoma Pueblo in New Mexico and you can buy a traditional, coil pot from one of the elder women. It will cost you, but it will be worth it. Or you could spend a lot less money on a knock-off. At Acoma, younger potters, many of whom are men, offer pots for sale that were molded, not coiled. What is remarkable from an anthropological standpoint is how the commercialization of pottery has changed gender relations. In societies worldwide, where pottery was made exclusively for domestic use, women were almost always the potters. Commercialization changed that, not only at Acoma Pueblo, but worldwide, as market economies transformed ancient ways of life. Changing gender roles in pottery making did not have to await the arrival of capitalist markets. Consider the case of Lapita pottery from Polynesia. The presumed ancestors of many Pacific cultures, people of Lapita culture began to migrate eastward across the Pacific Ocean at about 1500 BCE, reaching Tonga and Samoa by about 1000 BCE. Their pottery was a distinctive ware, decorated with repeating geometric patterns of dentate stamping. Anthropologist Yvonne Marshall believes that pots were made exclusively by women in traditional Lapita communities.[9] However, over time men got involved. Why? For Marshall, the answer traces to trade and ceremonialism. Ocean-going trade and its associated rituals were the purview of men, mostly. Increasingly, pottery production became geared towards non-domestic uses, with the more elaborate vessels funneled into exchanges controlled by men. Eventually, intensification of production broke down and ornate decoration disappeared. Throughout this period of change, plain pottery continued to be made by local communities (presumably women) for domestic uses. The Lapita case goes to show that any inducement to manufacture products beyond the level of domestic consumption introduces challenges to traditional operational sequences and their underlying social relationships. In this case, the expanding non-domestic "market" fueled demand for high-quality pots, while in the Acoma case it allowed for the development of cheap knock-offs. Despite the differences, both cases involved changes in gender roles and relations, reminding us of the impact that changing operational sequences can have on fundamental social dimensions.

Taking the long view on the history of traditional ceramics, many innovations effectively met short-term goals, and introduced unintended consequences and some new problems. The first kilns helped to concentrate heat, but they had higher construction costs and required more specialized fuels than open-pit fires. (Kilns required hardwood and other slow-burning fuels to attain and maintain temperatures in excess of 1,200 degrees Celsius). These costs were potentially offset by the longer use-lives of pots (if used in thermal applications, like cooking), but that would have dampened demand for new pots and thus thwarted the investment return of expensive infrastructure, notably the kiln itself. Over time innovations in kiln technology appeared—such as the Chinese climbing kiln (Figure 3.9), which optimized conduction—in some cases to address production demand, in others to decrease fuel costs. Clearly a major limit to production expansion for technologies involving enormous thermal energy was fuel. Until coal was introduced in the 19th-century Japanese ceramics industry, fuel consisted of wood or dung. Potting industries worldwide have contributed to local deforestation, and ultimately demand for more efficient kilns and alternative fuels, like coal and gas.

Unforeseen consequences beset the health of potters too, along with those who used their wares. Like their counterparts who manufactured gun flints centuries ago and unwittingly inhaled microscopic quartz, potters working with finely ground sand were prone to silicosis, a deadly lung disease. Likewise, British potters steeped in the tradition of lead glazing that fueled a worldwide demand for European tableware routinely suffered from lead poisoning. Add to this the collateral damage of lead exposure by consumers using pottery to process, store, and consume food. The lead threat continues today, showing how the production of something so traditional invites social interventions over labor rights, public health, and fair trade. Intensification always has its costs, direct and indirect.

Figure 3.9 Chinese climbing kilns like the one shown here date back as early as 5th century CE. The harnessing of the fact that heat rises was essential to the success of this innovation. [Photo by user Next-Exit (2007), shared under a CC-BY-SA 3.0 Unported License. Wikimedia Commons.]

Generating and Storing Energy

In our examples of flaked stone and pottery, matter was transformed through the application of mechanical energy, mostly controlled human force. In the case of flaked stone, force and know-how was used to reduce rock from larger to smaller sizes and from amorphous to formalized shapes. In the case of pottery, force and know-how was used to assemble a variety of substances and manipulate their respective properties to form various things, finish their surfaces, and harden them with fire. This last step goes beyond human force to involve thermal energy. Applied to either rock or clay, heat altered the physical properties of substances in ways useful to people. It may seem ironic that efforts to make rock more like glass were aimed at making its breakage more predictable, while efforts to make clay more glass-like were to lessen the risk of breakage while increasing thermal conductivity.

Ceramic Fuel Cells

Is it ironic or poetic that glass-like substances offer affordances by alternately breaking and not breaking? Take this to the microscopic level of transformation and we begin to understand that putting things together and taking them apart are two sides of the same coin. That is, at the level of physiochemical change—as in the process of vitrification—energy is absorbed, stored, and released in microcosmic versions of operational sequences. Change the conditions under which these transfers of energy occur and you can produce different outcomes and create different products.

Ceramic or solid oxide fuel cells are among the more promising products that offers an alternative to more traditional electricity generation. Consider their application in domestic uses of energy. When electricity is produced at a centralized power plant and distributed to homes via a grid, the electricity has to be used immediately and the source is often the combustion of coal or natural gas. However, a fuel cell that uses hydrogen and oxygen, for instance, produces power as it is needed, locally, in the home or a refrigeration truck. It does this by using a solid oxide membrane (electrolyte) to transport oxygen ions from one side of the fuel cell to the other. When these oxygen ions react with fuel such as hydrogen or carbon monoxide, electrons are released and one produces a current.

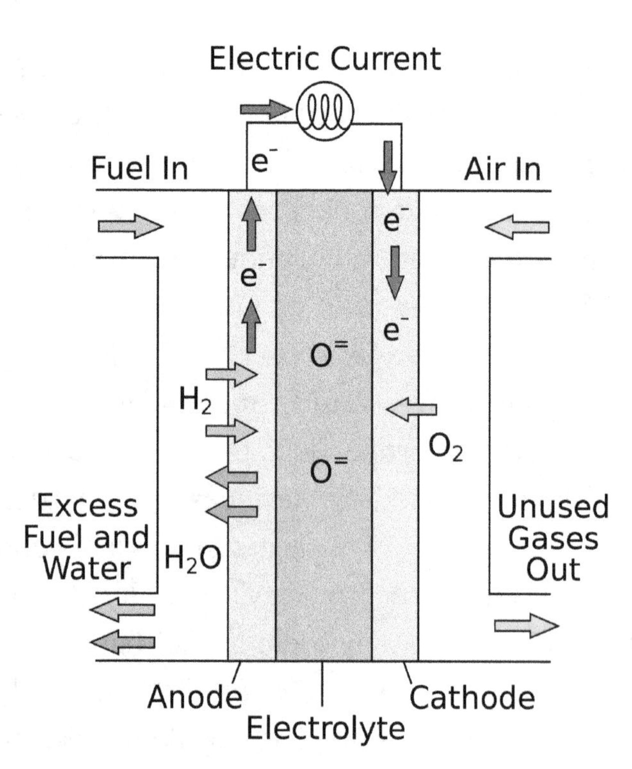

Figure 3.10 Solid oxide fuel cell. The electrolyte is typically a complex ceramic. [Wikimedia Commons.]

Two limitations of solid oxide fuel cells must be overcome before this technology gains a better foothold in the competition for energy production. These cells can be extremely efficient at producing electricity; however, they must operate at high temperatures in order to help with the production and transport of the oxygen ions. So the search continues for better solid oxide membrane materials to help reduce this operating temperature.

The second limitation is that fossil fuels are still involved in the process. Sure, they are not being combusted like they are in a conventional generator, but we still have the costs and impacts of creating the hydrogen of carbon monoxide fuel. What if, instead, fuel that is used to generate electricity could be generated from another source— say, the sun? Obviously, solar energy has been converted to electricity for decades now, and we all know the limitations of this technology, especially in places on the earth and at times of the year where sunlight is limited. In addition, storage of solar energy is a major limitation as it is much less efficient than the chemical storage of energy in the refined fossil fuels on which much of our energy-making infrastructure is based. So solid oxide fuel cells offer a potential way of converting solar energy to a chemical form that can be stored and used to create electricity when needed.

Getting the Most from the Sun

A promising innovation is the use of a ceramic substance, cerium dioxide, to convert solar energy into methane. The promise is nicely summarized in a short lecture by California Institute of Technology Professor Sossina Haile (https://www.youtube.com/watch?v=gSIsc7xBX3A). In this lecture, Professor Haile notes that conventional solar panels use only a portion of the spectrum of sunlight and thus fall short of their full potential for energy production. If we can capture and concentrate all the light, much like we do with a magnifying glass, we can take advantage of the full spectrum and produce lots of heat, which then can be used to drive the chemistry of fuel production. The "magic" of cerium dioxide, as Professor Haile sees it, is that it can quickly "breathe" oxygen at high temperatures. With nothing more than water, carbon dioxide, and solar heat, methane can be produced from the repetitive cycles of heating and cooling the ceramic surface. Not only is this sustainable from the standpoint of fuel production, but it helps to remove one of the greenhouse gases that is generated from combustion of fossil fuels.

Conclusion

Revealed in operational sequences are the sorts of entanglements between materials and society you read about in Gillespie's chapter on clay. In a sense, operational sequences offer methods for analyzing entanglements by showing how bodily and social acts are entangled with materials at every step of their extraction, processing, and refinement. This is no cookbook method any more than any operational sequence is merely technical steps in a production process. The broader historical, social, and cultural contexts of any process, at any point in time, is more than a backdrop for the materials engineering going on. These social and cultural factors offer rationale, precedent, contradiction, aesthetic, opportunity, creativity, value, and more. The manipulations of glass and ceramics involved a community as a whole, and in that sense we are all materials scientists.

We saw how the making of a Clovis point was entangled with the social structure of communal hunting, and how innovations in cooking pots were bound up in the emergence and growth of farming, and how the next generation of ceramic fuel cells are constrained by an existing infrastructure and social norms for fossil fuels, but capable of decentralizing locations of production and thus making households more self-sufficient. The intrinsic value of the analysis of operational sequences is insight on change that exists in the potential relationships between materials and society. In the cases described here, heat was a common medium for changing relationships between people and things. In all cases, heat enhanced or revealed affordances that were latent to the materials being manipulated by people. But the source of heat for realizing innovation had its limits and its hidden costs, as in the higher fuel costs of kiln firing or the environmental impacts of fossil fuel combustion. With such a long history of research and development, ceramic materials of the future will continue to offer alternatives to existing technologies. And with a history of impacting societies for millennia, the production and use of ceramic materials provide ample lessons for avoiding failures and enhancing our future successes.

Discussion Questions

1. Define the concept of *operational sequence* and explain how it encompasses social actions, as well as technical steps, in the making of things.
2. How is the manufacture and use of a stone tool like a Clovis point a social act? Even though Clovis points may have been made and used by men, what impact did the technology have on women?
3. What were the limitations of traditional subceramic cooking vessels in the application of sustained boiling and what did potters do to overcome those limitations?
4. What is *intensification* and how does it impact operational sequences? What does it mean to say there are *hidden costs* to intensification?

5. Do you see a promising future in ceramic fuel cells, and if so, what do you imagine to be the potential impact on society?

Key Terms

ceramic
operational sequence
cleavage plane
coiling
flaked stone
fluting
kiln
intensification
isotropic
subceramic
vitrification
chaîne opératoire
embedded procurement
amorphous
crystalline

Author Biography

Kenneth E. Sassaman is Hyatt and Cici Brown Professor of Florida Archaeology at the University of Florida. He earned a Ph.D. in Anthropology from the University of Massachusetts, Amherst in 1991 and has over thirty-five years of archaeological field experience in the indigenous history of the American Southeast. His most recent work centers on the challenges of climate change on the gulf coast of Florida over the past 5,000 years. He is the author of over one hundred articles and book chapters, and the author or editor of nine books, most recently *The Archaeology of Ancient North America* (with Tim Pauketat; Cambridge University Press, 2020).

Notes

1. Peter Bleed, “Trees or Chains, Links or Branches: Conceptual Alternatives for Consideration of Stone Tool Production and Other Sequential Activities,” *Journal of Archaeological Method and Theory* 8 (2001): 101–27, https://doi.org/10.1023/A:1009526016167.
2. John C. Whittaker, *Flintknapping: Making and Understanding Stone Tools* (Austin, TX: Univ. of Texas Press, 1994), 65-72, http://www.worldcat.org/oclc/1041322933.
3. Bruce A. Buck, “Ancient Technology in Contemporary Surgery,” *Western Journal of Medicine* 136, no. 3 (1982): 265–69, https://www.ncbi.nlm.nih.gov/pmc/articles/PMC1273673/.
4. Gary Haynes, *The Early Settlement of North America: The Clovis Era* (Cambridge: Cambridge Univ. Press, 2002), http://www.worldcat.org/oclc/49327000.
5. Charlotte Beck, Amanda K. Taylor, George T. Jones, Cynthia M. Fadem, Caitlyn R. Cook, and Sara A. Millward, “Rocks are Heavy: Transport Costs and Paleolithic Quarry Behavior in the Great Basin,” *Journal of Anthropological Archaeology* 21, no. 4 (2002): 481–507, https://doi.org/10.1016/S0278-4165(02)00007-7.
6. Lewis R. Binford, “Organization and Formation Processes: Looking at Curated Technologies,” *Journal of Anthropological Research* 35, no. 3 (1979): 255–73, https://doi.org/10.1086/jar.35.3.3629902.
7. Brian S. Robinson, Jennifer C. Ort, William A. Eldridge, Adrian L. Burke, and Bertrand G. Pelletier, “Paleoindian Aggregation and Social Context at Bull Brook,” *American Antiquity* 74 (2009): 423–47, https://doi.org/10.1017/S0002731600048691.
8. David P. Braun, “Pots as Tools,” in *Archaeological Hammers and Theories*, eds. J. A. Moore and A. S. Keene (New York: Academic Press, 1983), 108–34, https://doi.org/10.1016/C2013-0-11196-8.
9. Yvonne Marshall, “Who Made the Lapita Pots?: A Case Study in Gender Archaeology," *Journal of the Polynesian Society* 94, no. 3 (1985): 205–33, http://www.jstor.com/stable/20705934.

Concrete: Engineering Society through Social Spaces

MARY ANN EAVERLY

> "To be ignorant of what occurred before you were born is to remain always a child. For what is the worth of human life unless it is woven into the life of our ancestors by the records of history?" —Cicero

Abstract

The most popular building material in the world today, concrete was first developed and exploited by the ancient Romans, who used it to create monumental public spaces such as **aqueducts**, churches, baths, and the Colosseum. Although concrete is a durable material with many building applications, the design of Roman concrete structures reinforced Roman ideas about social status and imperial power. This chapter explores the rich history of concrete and its legacy in the modern world, touching upon the role of concrete in ancient Rome, today's technical advances in concrete construction, and concrete's environmental drawbacks. The chapter also examines how concrete construction is shaped by societal ideals today, just as it was by societal ideas in ancient Rome.

Introduction

Pourable, moldable, durable, waterproof, and relatively easy and inexpensive to manufacture, concrete is the world's most popular building material. We live, work, and play on and in buildings and roads constructed from it. Architects exploit its properties to create artistic tours de force as well as utilitarian monuments (Figure 4.1).

Concrete is such a part of our daily lives that we may not stop to think about who invented it or why builders create certain types of buildings from it and not others. Is there a connection between buildings and larger societal forces? Have you ever wondered why we use concrete the way we do? Do buildings reflect a society's ideals for social organization? To try to answer these questions we need to examine the role concrete played in the society that first developed it—ancient Rome during the late Roman Republic era (200–100 BCE). This chapter explores how the use of concrete in a society has historically been connected to the forms of social organization that structure that society.

Figure 4.1 The Guggenheim Museum, New York City [Photo by Jean-Christophe Benoist (2012), shared under a CC BY 3.0 Unported License. Wikimedia Commons.]

Origin of Concrete

As far back as the sixth millennium (6000–5000 BCE), the ancient Mesopotamians knew that heating calcium carbonate, a substance occurring naturally in limestone rocks, creates a new substance, known today as quicklime, in a process described chemically as $CACO_3$ + heat(1000° C) =CO_2 + CaO. This chemical reaction releases carbon dioxide into the atmosphere (more on this later). The resulting material, when mixed with water, bonds to other surfaces. The early residents of Çatalhöyük, an ancient city in modern-day Turkey, as explored in Gillespie's chapter on clay, used this substance to coat their walls, providing a surface for painted decoration.

Mount Vesuvius and Pompeii

In 79 CE, long after the architect Vitruvius's death, Mount Vesuvius erupted and destroyed the Roman city of Pompeii, preserving its buildings and construction practices (as well as many of its human inhabitants and even a few dogs) for modern archaeologists to study. This greatly enhanced our understanding of early concrete. Find out more from Cyark.

Figure 4.2 Pozzulana from the Area of Vesuvius [Shared under a CC BY-SA 2.5 Generic License. Wikimedia Commons.]

The Egyptians of the third millennium (3000–2000 BCE) used quicklime as mortar to join stones in stone construction.[1] In both of these cases, concrete was simply joined to other materials as decoration or adhesive; it was not yet a primary building material. However, the Romans in the 3rd century BCE discovered that by mixing quicklime and sand with a local volcanic stone—**pozzulana**—they could create something much stronger and more durable than simple quicklime (Figure 4.2). The Romans called their new material ***opus caementicum***; it is the forerunner of modern concrete.

Because of its durability, many ancient Roman buildings stand today, over 2,000 years later, providing evidence of the strength of Roman concrete and Roman construction practices. Questions of materials' durability also are important when we talk about recording and storing information (see Effros, "Writing Materials"). In addition to the physical structures, we have ancient testimony concerning Roman buildings from the Roman architect Vitruvius, who wrote during the reign of the emperor Augustus (27 BCE–14 CE). Vitruvius wrote a 10-volume history of Roman architecture entitled *De Architectura*. An architect himself and a former catapult operator (catapults hurled projectiles at walls during sieges) in the Roman army, he was interested in many of the practical aspects of Roman construction. In the second volume of his history, Vitruvius describes several building materials, including concrete. About pozzulana, the additive that makes Roman concrete possible, he says, "There is also a type of powder that brings about marvelous things naturally. It occurs in the region of Baiae and in the countryside that belongs to the towns around Mount Vesuvius" (*De Architectura*, II, 6, 1) [2] Since Italy is a volcanic region, pozzulana was easy to find.

While *opus caementicum* had many of the same properties as modern concrete, it was strong, moldable, and lighter in weight than stone and lacked the smooth, pourable consistency of concrete today. In addition to pozzulana, the mix included aggregate—rubble, pieces of rough stone, and broken brick. These materials were not blended into one seamless product, but were instead bound in a rough mass. The resulting mixture then had to be laid by hand rather than poured. The Romans found its rough appearance unsightly and covered or faced it with other materials, usually a surface layer or veneer of brick or fine marble. Even with the added veneer, concrete construction proved much more economical than the previous reliance on stone for large-scale buildings. Stone carving required highly skilled laborers. Transporting blocks from the quarry to the building site was time consuming and expensive, as we shall see later in this chapter. Concrete, in contrast, could be made on site and laid by less-skilled workers, who could be organized into quick-working groups.[3] Covering buildings with a veneer of fine marble or other exotic stones, the Romans achieved the appearance of an expensive solid stone structure more efficiently and with less cost than they would have for an all-stone building.

Roman Concrete Revolution

Concrete freed the Romans from the constraints of traditional architecture. Before concrete, buildings made of wood or stone used what was known as the **post-and-lintel** construction system, in which vertical elements (posts) support horizontal elements (lintels). You can see post-and-lintel construction in the columns of such famous ancient Greek temples as the Parthenon in Athens[4] (Figure 4.3). With this system, it is almost impossible to create a large, unsupported, roofed space. While some Greek temples were massive—the Temple of Apollo at Didyma in modern Turkey had a total area of 18,000 square feet and columns that were 64 feet tall— their usable interior space was limited because of the need for internal roof supports (columns or posts), which took up much floor space.[5]

Figure 4.3 Post-and-Lintel Construction. The Parthenon, Temple to Athena, 5th century BCE, Athens, Greece [Photo by Tim Bekaert (2005), shared in the Public Domain. Wikimedia Commons.]

Figure 4.4 The Pantheon, Rome (2nd century CE) [Photo by Maros Mraz (2008), shared under a CC-BY-SA 3.0 Unported License. Wikimedia Commons.]

Because of their light weight and moldability, concrete roofing systems did not need to be supported. By using concrete to create intersecting arches and vaults, the Romans designed interior spaces on a far grander scale than post-and-lintel construction allowed. Archaeologists called the Roman exploitation of concrete arches, vaults, and domes to create interior space the **"concrete revolution."** Among the most dramatic of these buildings is the Pantheon in Rome (Figure 4.4).

The Pantheon

From the exterior, the Pantheon looks like a traditional temple with columns for support, but the interior is a spectacular domed space. Compare its interior with that of the Parthenon, considered the most perfect of Greek temples (Figures 4.5 and 4.6).

Figure 4.5 The Pantheon Interior, Rome (2nd century CE) [Photo by Stefan Bauer (2005), shared under a CC BY 2.5 Generic License. Wikimedia Commons.]

Because the diameter and the height of the Pantheon are the same, the building's interior encloses a complete sphere of space, which may be an allusion to the totality of the gods (the word *pantheon*, derived from ancient Greek, means "all of the gods") since a circle is a complete form.[6] While columns built into the walls appear to support the ceiling, concrete arches and vaults actually bear the weight. By using increasingly light materials in the concrete aggregate as they moved to the top of the dome, the builders were able to ensure that it did not collapse. Until the 21st century, the Pantheon was the largest unsupported dome in the world. All of the structural, load-bearing work of the concrete is hidden beneath elaborately colored marble veneer. Note that despite its "revolutionary" interior, the Romans gave the building the outward appearance of traditional post-and-lintel building style. They admired the architectural achievements of the Greeks, so they retained the column styles that the Greeks had created (Doric, Ionic, and Corinthian), even when these columns had no true structural role in a building. This tradition continues today in modern buildings, such as banks or government offices, in which ancient Greek architectural elements on concrete buildings evoke the perceived glory of the Classical past.

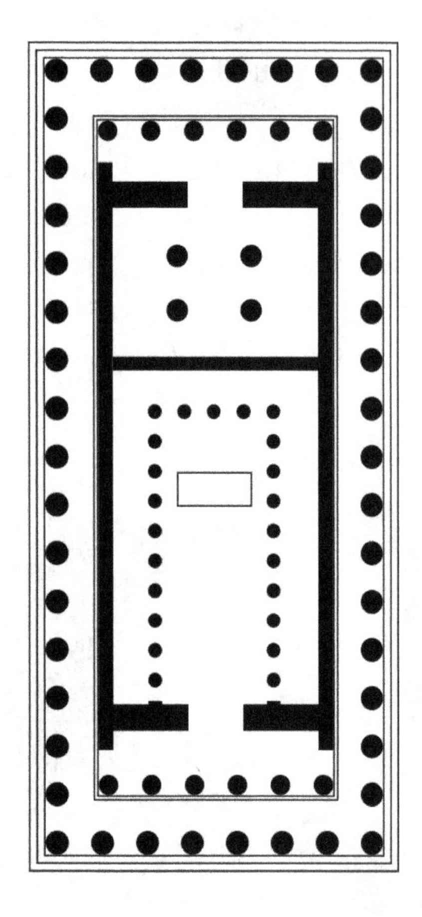

Figure 4.6 Plan of the Interior of the Parthenon. Each dot represents a column. [Image by user Argento (2006), shared in the Public Domain. Wikimedia Commons.]

The exterior columns of the Pantheon were not, as is typical in ancient temples, made from segments, but were instead each carved from a single block of granite brought to Rome from Egypt. Costly and difficult to carve and transport, they, along with the exotic marble veneers inside the building, emphasized the reach and power of the Roman Empire. Using concrete cut down considerably on construction time. Including the time to transport the columns from Egypt, the Pantheon took just six years to complete (118–125 CE), while the all-stone Greek temple of Apollo at Didyma took almost 500 years (332 BCE–130 CE) to complete.

Bread and Circuses

As impressive an engineering feat as the Pantheon is, temples did not provide the impetus for Roman exploitation of concrete. Instead, spectacles, bath complexes, and military operations drove the development of this technology. As archaeologist Lynne Lancaster writes: "The two cultural institutions that had the greatest effect on the advances in vaulted technology during the imperial period were public bathing and public entertainment."[7] Ancient Roman writers attest to the importance of these cultural practices. In the 2nd century CE, the Roman writer Juvenal, despairing of what he perceived as the decline of the Roman national character, states in *Satires*:

> The people that once bestowed commands, consulships, legions and all else, now meddles no more and longs eagerly for just two things—Bread and Games. (*Satires*, 10.81) [8]

This phrase, usually translated as "bread and circuses," refers to the government-sponsored daily distribution of free bread to the populace and spectacles such as gladiatorial games that took place in **amphitheaters**, large concrete structures designed to house such events. What role did these games play in Roman society, and how did they contribute to the rise of concrete technology?

The Colosseum and Roman Gladiatorial Games

The most enduring statement of the Roman love of spectacle is the Colosseum in Rome (Figure 4.7). Designed to hold 50,000 spectators, the building covers an area of 615 feet by 512 feet, or almost 315,000 square feet, and is 159 feet tall. Eighty exits facilitated easy entrance and exit. The building fully exploits the properties of concrete. While from the exterior it looks as if columns, a feature of post-and-lintel construction, are doing the supporting work, they are simply a façade (a decorative surface) covering the actual structural elements—concrete arches and vaults that support the sloped (stadium) seating. These arches and vaults allowed the Romans, unlike the Greeks, to build stadium seating in the round. While the Greeks had developed stadium seating for their theaters, they needed a hillside to provide the inclined angle for the seats. Concrete obviated the need for a naturally occurring slope.[9]

Figure 4.7 Drawing of the Colosseum, Rome. In the cut-away sections, note the use of vaults and arches for support. [Drawing by Jaakko Luttinen (2012), shared under a CC-BY-SA 3.0 Unported License. Wikimedia Commons.]

Activity: Take a video tour

This aerial video of the Colosseum (https://www.youtube.com/watch?v=B4GvcCWzZZg) offers a bird's-eye view of the exterior and interior, with a detailed look at the concrete arches and vaults that supported stadium seating.

Figure 4.8 Pollice Verso, by Jean-Léon Gérôme (1872). This 19th-century painting attempts to recreate the moment of death for a defeated gladiator. The crowd is condemning him to death by pointing thumbs down. We do not, however, actually know what gesture was used to determine life or death. [Phoenix Art Museum. Wikimedia Commons.]

Begun in 72 CE and completed in 80 CE, this amphitheater hosted one of the favorite forms of Roman entertainment—gladiatorial games (Figure 4.8). The games included not only man-to-man combat, but also a daylong program comprised of public executions, man-vs.-beast and animal-vs.-animal contests, and even mock naval battles in which participants fought to the death. While scholars debate the origin of these blood sports, most believe that they derive from early funeral rituals involving games and blood sacrifice offered to appease the spirits of the dead.

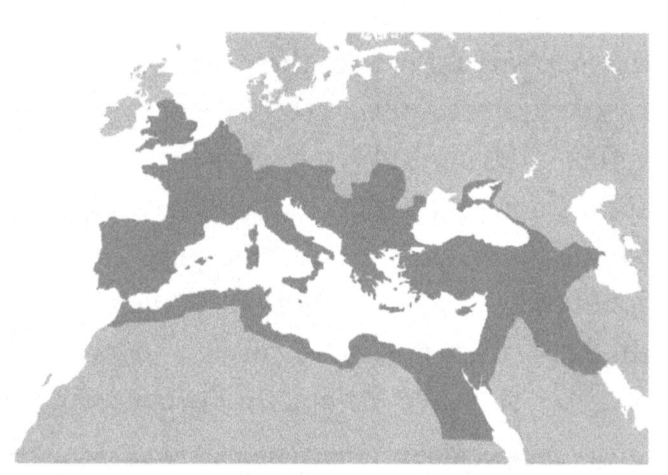

Figure 4.9 Map of the Roman Empire at its greatest extent. The darkest areas (green) on this map was controlled by the Romans. [Map by user Angelus (2011), shared under a CC-BY-SA 3.0 Unported License. Wikimedia Commons.]

By the time of the Colosseum, however, these activities were intended to reinforce social and imperial identity throughout the empire. How did this work? First, where one sat was determined by one's social class. Roman society had four sharply delineated main divisions. The patricians—hereditary noble families—formed the top (aristocratic) class. The most important member of this group was, of course, the Emperor, absolute ruler of a vast empire (Figure 4.9). Next in importance were the plebeians—the majority of Roman citizens (free-born but not patrician). They were followed by freedmen—individuals who had once been enslaved but who had, through various means, earned their freedom. Finally, enslaved people formed the lowest level. This group included prisoners captured in Rome's many wars throughout the ancient Mediterranean, but also included Romans who had been sold into slavery because of debt and those who had been born to enslaved parents. The best seats, in the Colosseum and other Roman amphitheaters, near the arena floor and thus closest to the action, belonged to the patricians. The remainder of the spectators sat in descending order of importance (plebeians, freedmen, and slaves). The higher the seats, the lower the status. Women—who were not eligible to vote or to participate in government—sat at the very top with enslaved individuals, indicating their inferior status regardless of social class. Amphitheaters were a vital part of every Roman town and this arrangement was repeated throughout the empire.

Figure 4.10 "The Swamp." Ben Hill Griffin Stadium, University of Florida, Gainesville [Photo by Douglas Green (2005), shared under a CC BY 2.0 Generic License. Wikimedia Commons.]

Gathered to watch the games, Romans could, while surveying the audience, reaffirm their own place in society. Seating provided physical and visual confirmation of society's rules. Compare the Colosseum with our modern stadiums. If we look at an American college football stadium during a Saturday game, we see that some status functions are also at work, although in this case we see that our seating reverses Roman practice. In American football stadiums the most expensive seats, and therefore the seats belonging to those who might be called "the most important people," are the skyboxes at the *top* of the stadium (Figure 4.10). These are enclosed and provide food, beverages, and the best view of the game. Yet, in at least one respect, the Colosseum had innovations lacking in the modern world. While only a few American football stadiums are domed, the Colosseum and many local Roman amphitheaters provided a retractable awning called a *velarium* to protect the audience from the sun. American professional basketball arenas, where the seats do follow the Colosseum pattern, also inform us about status in the modern world. In this case, the "patricians," seated closest to the action, are those whom our society seems to value most—celebrities from sports, film, and recording industries.

Key Concept: Gladiators

Highly trained combatants wearing different types of specialized armor fought against each other in single combat. For example, a heavily armored man with a sword (Myrmillo) would fight a lightly armored man with a net and three-pronged pitchfork (Retiarius). The net man had the advantage of speed and the armored man the advantage of protection. The Romans enjoyed seeing each man exercise his skill against the other. The loser was typically killed by his opponent, although if he had fought well he might be spared by the administrator of the games to fight again. The combat always paused for the moment when the loser confronted his own death. The crowd admired those who faced their deaths bravely.

The Colosseum performed an additional societal function. Showcasing creatures from all parts of the ancient world (elephants from Africa, tigers from India, etc.) in the animal combats, the emperor showed people the extent of the empire and the power of an emperor able to control such a vast territory. While acknowledging the emperor's authority, they could take pride in belonging to a society that had seemingly mastered the entire world, as they knew it. The message of unity presented at the Colosseum also finds echoes in the modern American college football stadium. Students and alumni gather to confirm their identity and unity as members of a collegiate community despite different majors, academic programs, and class years. Professional football teams can unite the disparate members of a city as well when fans fill the stadium on a Sunday to cheer their team. World Cup Soccer fever shows the intensity of the connection between athletics and national pride today. Concrete stadiums and arenas continue to reinforce social ideals.

The Colosseum also served as propaganda supporting Emperor Vespasian. It was a public building placed over the demolished remains of the private villa of Vespasian's hated predecessor, Nero. The land had originally been the site of Roman private homes. Nero had confiscated it for his own private pleasure palace after a tragic fire destroyed much of Rome in 64 CE. The building takes its name not from its size, but from its proximity to a colossal statue of Nero, in the guise of the sun-god, which Vespasian left standing. The contrast between the two monuments, a public place of entertainment and an ego-enhancing statue that once decorated a private luxury palace, provided a continued reinforcement of Vespasian's message of benevolence toward the people of Rome.

Mock naval battles, *naumachia*, in the Colosseum provided another affirmation of the power of the emperor. Ancient sources tell us that the Colosseum floor could in fact be flooded (remember the waterproof nature of concrete). These battles did not reenact contemporary Roman victories, but instead depicted battles from the past. Choosing historical battles allowed the emperor to show that he had control not only over the physical terrain of his empire, but also over time.

Water Supply

The naumachia were supplied with water by aqueducts, another feature of Roman engineering connected to concrete. An aqueduct is a water transport system. To bring water from its source, a spring or lake, over long distances required keeping the water constantly flowing. Lacking modern electrical pumps, the Romans relied on raising and lowering the water's level (Figure 4.11). Aqueducts, carried on arched stone or concrete substructures, spanned valleys and other topographical obstacles to keep water moving. They supplied an enormous volume of water to the city. At the height of Rome's population of one million people, eleven aqueducts supplied the city with the equivalent of 540 liters, or about 142 gallons, of water per person per day.[10] Reading this, one might think that every Roman had running water at home, but that was not the case. Though water did flow at public fountains, the Romans primarily used piped water for lavish fountain displays in gardens in the homes of the wealthy. An example of this is seen at the home of an aristocrat in Pompeii, in the fountain remains preserved by the eruption of Vesuvius. These aristocrats proclaimed their power and wealth by having purely decorative water displays in their homes (Figure 4.12).

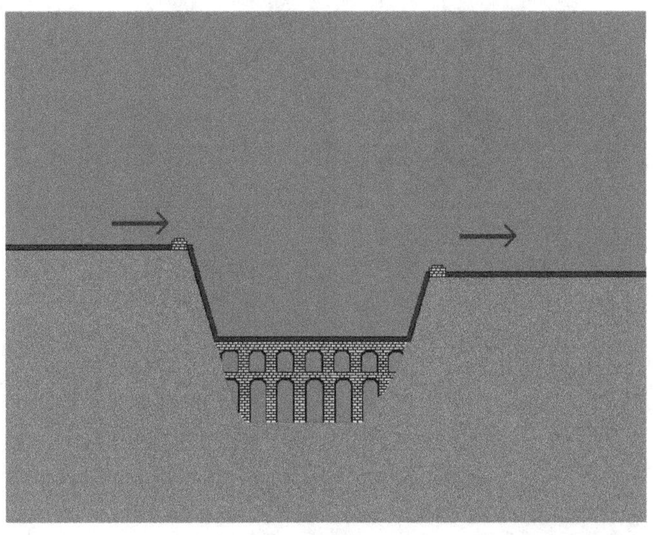

Figure 4.11 Diagram showing how aqueducts keep water flowing across obstacles such as valleys. The angle of descent was carefully calculated so that the water would continue moving without pumps. [Wikimedia Commons.]

Figure 4.12 Water Feature from the Gardens of Loreius Tiburtinus, Pompeii, 1st century CE. Water flowed through this channel to create a man-made river in this aristocratic home. [Photo by user Magistermercator (2010), shared under a CC BY-SA 3.0 Unported License. Wikimedia Commons.]

While such display was important to promote status, an important function of the aqueducts was supplying water to the public bathhouses, another major factor in the development of Roman concrete technology. One of the earliest domed concrete structures in the Roman world is a 2nd-century-BCE bath complex in the Italian city of Baiae. The need for water was two-fold. Baths, which exploited concrete in their construction, needed water for bathing as well as for manufacturing the concrete. Once again, societal needs for public spaces drive concrete construction.

For the Romans, bathing in public bath houses was a vital daily activity. A 1st-century-BCE Roman felt that such baths were so important that he inscribed the following on his tombstone: "Wine, sex, and baths ruin our bodies, but they are the stuff of life."[11] Every city had numerous baths. The bathing process followed a progression through baths of different temperatures, from cool to hot and back to cool. Patrons could also exercise, buy snacks, and partake of beauty treatments such as hair plucking, activities similar to those found in modern health-and-fitness centers. The hottest rooms were heated by a **hypocaust** system—raised floors and tile pipes—that allowed air heated by furnaces beneath the floors to rise through pipes in the walls. The air was forced from the furnaces by bellows, hand-pumped devices that produced a strong current of air when squeezed. The addition of water in some rooms created steam. There were also unheated pools for plunging. The fact that concrete is waterproof made it ideal for bath buildings. Not only did baths provide for public cleanliness, but they also functioned as a social safety valve. While the Colosseum, and life in general, emphasized class distinctions, the baths allowed for a temporary dissolution of those same social levels. Everyone, regardless of social class, "got naked together" (albeit with men in one section and women in another section or in separate facilities).

These buildings were designed to be luxurious and a large domed hot room, called a caldarium, became a standard feature. Beginning with the emperor Titus in the first century BCE, emperors sponsored lavish imperial bath buildings in Rome. Such elaborate constructions curried favor with an often restless population by emphasizing the emperor's benevolence (or seeming benevolence, since many emperors were violent, unstable individuals), while highlighting the power and expanse of the empire he controlled. Like the Pantheon, these buildings often used expensive and exotic marble veneer to cover the concrete. Succeeding emperors tried to outdo each other by building larger and larger complexes. One of the grandest of these, portions of which still stand, was constructed during the reign of the emperor Caracalla in the 2nd century CE. Elaborately decorated with colossal mythological statues, it enclosed an area 1,315 feet by 1,076 feet. Its size and grandeur continued the tradition of imperial display and provided a needed social outlet for Rome's populace (Figure 4.13). Even the poorest Roman could briefly enjoy beautiful—and, in winter, warm—surroundings thanks to the emperor.

Figure 4.13 Baths of Caracalla, Rome, 2nd century CE. Note the fragments of mosaic decoration. [Photo by user Karelj (2011), shared in the Public Domain. Wikimedia Commons.]

Military expansion and trade made possible the vast empire celebrated in imperial buildings such as the Colosseum and the Baths of Caracalla. Concrete was also exploited for these goals. Because concrete could set underwater, it was perfect for creating ports that needed strong underwater substructures. Vitruvius does not fail to mention this important characteristic of the composition of *opus caementicum*:

> Hence, when these three ingredients (lime, fired rubble, and pozzolana), forged in similar fashion, by fire's intensity, meet in a single mixture, and put into contact with water the ingredients cling together as one and, stiffened by water, quickly solidify. Neither waves nor the force of water can dissolve them. (*De Architectura*, II, 6, 1)[12]

Figure 4.14 Traces of Roman Port at Hersonisos, Crete [Photo by Jose Mario Pires (2009), shared under a CC BY-SA 3.0 Unported License. Wikimedia Commons.]

Thus secure, deep pilings could be put in place for ports throughout the Roman world. Concrete allowed the Romans to spread their military forces and to develop trade routes by creating ports in areas that did not naturally have adequate facilities. Ports provided part of the infrastructure for expanding, consolidating, and ruling the vast territories that comprised the Roman Empire. Along with baths and amphitheaters, these man-made harbors reflect the societal values that dominated the development and exploitation of concrete's inherent properties (Figure 4.14).[13]

The emperor's desire for control is evident in the Colosseum and imperial bath complexes. Social class is emphasized in the Colosseum and social tensions are relaxed in the baths. By creating concrete ports, Romans controlled distant territories through transport of troops for military campaigns and movement of goods through trade. They were thus able to spread their values and social organization throughout the ancient Mediterranean world. Their particular worldview shaped the ways in which concrete could be used. While we might expect other types of monuments—durable public housing, perhaps, or water delivered directly to every home—these were not important to the Romans and thus were not among the reasons for their development of concrete forms. The Colosseum, baths, and port facilities promoted Roman identity and power. Although the Roman Empire eventually collapsed, its concrete structures endure. In fact, modern engineers and archaeologists are studying Roman port construction to see if these installations can teach us something about durability.

Modern Concrete

Figure 4.15 Soldier Field, Chicago, Illinois. Note the combination of modern concrete and classical columns. [Photo by J. Crocker (2010). Wikimedia Commons.]

We have seen how concrete technology was driven in certain directions by Roman social forces. What about us? What societal factors determine concrete's use today? We use concrete for many types of buildings and infrastructure—museums, houses, and bridges, for example. Some of these remain much the same in style and form as those the Romans created. For example, modern athletic complexes are startlingly similar in shape and seat arrangement to ancient amphitheaters (Figure 4.15). Although our athletes do not literally fight to the death, we continue to place a high value on athletic competition and on venues designed to showcase it. Universities and cities pride themselves on their multimillion-dollar sports complexes. Professional teams can threaten to leave cities if taxpayers do not fund updates to their stadiums or new stadiums. Modern consumer culture drives other types of concrete structures, such as shopping malls. Transportation needs promote airport construction. Concrete continues to be, as it was for the Romans, a relatively inexpensive and timesaving material. Yet, in the modern world, it has tremendous hidden costs.

Problems and Challenges

Despite more highly mechanized manufacturing techniques, the process of creating concrete still relies on the basic chemical reaction exploited by the Romans, the reaction that releases carbon dioxide, a major pollutant, into the atmosphere. The absence of many other man-made pollutants in ancient times and the Romans' smaller manufacturing scale meant that they did not suffer from pollution to the same degree we do. The world's yearly production of concrete has been rising steadily and today is over four billion tons. Concrete production is responsible for nearly eight percent of the anthropogenic (human-made) greenhouse gases released into the air. By some estimates, the external climate and health damages caused by concrete production amount to approximately 74 percent of the value of the industry itself, putting the external costs of concrete higher than natural gas and oil and only slightly less than coal.[14]. As even more concrete is produced, the amount of carbon dioxide will also rise unless we develop smarter, greener methods of production. Concrete manufacturers recognize this problem, but any solution will have to be cost effective for worldwide adoption to take place.

An additional cost of manufacturing concrete is the need for sand as a component of the finished product. Today, sand is becoming an increasingly rare and sought-after commodity. As the Romans knew, only sand worn by water (river or ocean sand), not sand that has been exposed to the elements (desert sand), is suitable. As Vitruvius states:

> When sand beds lie exposed for any stretch of time after they have been worked, subjected to sun and moon and frost, they break down and become earthy. And thus when such sands are mixed into the mortar, they cannot hold the rubble together. Instead, the rubble comes loose, and the weight of the masonry, which the walls can no longer sustain, collapses. (*De Architectura*, II, 4, 3)[15]

The vast deserts of the world cannot supply the right kind of sand. Excessive removal of river and sea sand is already destroying fragile ecosystems. Modern battles over this dwindling resource have resulted in murder in some parts of the world. Residents opposed to sand mining in one community in India were killed by groups controlling the manufacture of concrete for shopping malls and stadiums.[16]

While these problems are connected with new construction, older concrete structures pose other issues. Most modern concrete is reinforced by metal bars that lead to eventual cracking as the metal expands and contracts. Can we recycle ruined concrete buildings? How can we stabilize and repair buildings? Engineers are working to develop new technologies that can sense imminent structural issues before a bridge or building collapses. To prevent damage in new construction, engineers developed Smartcrete, a form of concrete that can repair itself. New methods of concrete construction such as Ductal, which requires no metal, and the use of cloth as a framing material are also potential answers to this problem.

Although some striking modern architectural monuments have been built from concrete—the Guggenheim Museum in New York, for example—many consider modern concrete stark and ugly because of the many utilitarian buildings constructed from it. Because, unlike the Romans, we can make concrete with a smooth surface, it is not necessary for us to cover it with other materials. After World War II, when a devastated Europe was in need of quick and cheap housing, architects turned toward concrete.[17] While answering a key societal need and advancing the idea of an affordable and equal form of housing for everyone, they filled cities with identical, unappealing structures.

Figure 4.16 Musée des Civilisations de l'Europe et de la Méditerranée, Marseille, France. Note the use of Ductal for the lacy lattice work. [Photo by Jean-Pierre Dalbéra (2014), shared under a CC BY 2.0 Generic License. Wikimedia Commons.]

Liquid Stone: New Architecture in Concrete, an exhibition held in 2004–2006 at the National Building Museum in Washington, DC, identified a new design direction in concrete architecture, a movement toward more dramatic and aesthetically pleasing buildings. Using fabric to mold concrete, creating translucent concrete, and embedding fiber-optics in concrete all create dramatic new visual effects. In addition, newer forms of concrete can create sculptural embellishments for buildings at a fraction of the cost of stone (Figure 4.16).[18] New concrete technologies continue to emerge. Among the possibilities is concrete laid by robots.[19] Acknowledging the close link between buildings and social structure allows us to wonder if there might be hidden costs to this technology. What types of workforce changes would occur if machines took over this aspect of building construction? Would using robots free humans to do other things or would it merely eliminate a large category of jobs?

Future of Concrete

Concrete's connection to social organization continues. In 2015, it was reported that a group of environmentalists, marine biologists, and nautical engineers were designing a floating city (seastead) on concrete piers, with plans for 300 people focused on examining pressing world problems such as hunger and health issues to inhabit the city. As of 2020, the idea continues to be explored. Legal standing and sovereignty of the proposed settlements have been thorny issues for the proposed projects.

Building construction remains the primary use of concrete today. Like the Romans, we limit concrete to certain types of applications that fit our society's needs. We build apartments, shopping malls, stadiums, and airports.

Might there not be other uses for such a versatile material beyond architecture? Architects, engineers, and others are beginning to address this question. For example, kitchen designers are using concrete for countertops, taking advantage of its durability, cost effectiveness, modern appearance, and ability to resist water. People are even considering how ancient Roman concrete might be used to address sea level rise associated with global warming. [20]

To encourage thinking about a common material in a different light, the American Society for Civil Engineering sponsors an annual concrete canoe contest. This challenge forces students to broaden their ideas about possible applications for this common material. Engineering students from across the US attempt to build and race a concrete canoe. They are judged not only on the results of the race, but also on their design concept. Concrete is certainly not the first material that comes to mind when thinking about canoes, although it is waterproof. But a concrete canoe suggests that if we think beyond the limits imposed on the use of concrete by our societal worldview and historical traditions, we may be able to find newer and more effective ways to use this versatile material.

Activity: Learn about the Concrete Canoe Competition

The 2012 competition, hosted by the University of Nebraska-Lincoln. [Photo by Missouri S&T Student Design & Experiential Learning Center, shared under a CC BY-ND 2.0 Generic License. Flickr.]

As shown in this video (https://www.youtube.com/watch?v=au0dTOridug&t=39s), the annual Concrete Canoe Competition offers students "an opportunity to gain hands-on, practical experience and leadership skills by working with concrete mix designs" as they build and race their canoes. Hosted by the American Society of Civil Engineers, the competition dates back to the 1980s.

Conclusions

Concrete has been used for thousands of years and has enabled many architectural revolutions starting around the time of the Romans. Today it remains the most used material in society by volume and weight. There is no doubt concrete will continue to be used extensively. However given the concern for the adverse environmental impacts of making concrete, it's clear that future concrete innovations will require not only creative applications but also creative manufacturing methods to help mitigate its carbon footprint.

Discussion Questions

1. What types of construction were the driving forces behind Roman concrete construction?
2. What societal needs did these building types reflect?
3. What was the composition of Roman concrete and how does it differ from modern concrete?
4. What new uses can you think of for concrete?
5. Is there anything in its inherent properties that limits it to current uses or are other avenues waiting to be explored?
6. What additional societal needs could concrete fill?
7. If it cannot be produced more cleanly, does its environmental impact mean that concrete is not worth the cost?

Key Terms

aqueduct
amphitheater
concrete revolution
hypocaust
opus caementicum
post-and-lintel construction
pozzulana

Author Biography

Mary Ann Eaverly, Professor and Chair of the Classics Department at the University of Florida, received her AB in Classical and Near Eastern Archaeology from Bryn Mawr College and her PhD in Classical Art and Archaeology from the University of Michigan. She was the Vanderpool Fellow at the American School of Classical Studies in Athens. Among her publications are *Archaic Greek Equestrian Sculpture* (Univ. of Michigan Press, 1995) and *Tan Men/Pale Women: Color and Gender in Archaic Greece and Egypt, a Comparative Approach* (Univ. of Michigan Press, 2013) She is also interested in the use of mythological and archaeological imagery in the work of modernist women poets and has co-authored several articles on this topic with Marsha Bryant (UF Department of English), including most recently "Modernist Migrations, Pedagogical Arenas: Translating Modernist Reception in the Classroom and Gallery," in *The Classics in Modernist Translation*, ed. Lynn Kozak and Miranda Hickman (London: Bloomsbury Academic, 2019).

Notes

1. Jean-Pierre Adam, *Roman Building: Materials and Techniques*, trans. Anthony Matthews (Bloomington: Univ. of Indiana Press, 1994), 65, http://www.worldcat.org/oclc/54913642.
2. Ingrid D. Rowland, trans., *Vitruvius: Ten Books on Architecture* (New York: Cambridge Univ. Press, 1999), 37–38, http://www.worldcat.org/oclc/779894769.
3. On this point see Lynne C. Lancaster, *Concrete Vaulted Construction in Imperial Rome: Innovations in Context* (Cambridge: Cambridge Univ. Press, 2005), 18–20, https://doi.org/10.1017/CBO9780511610516.
4. Judith M. Barringer, *The Art and Archaeology of Ancient Greece* (Cambridge: Cambridge Univ. Press, 2014), 225–240, https://doi.org/10.1017/CBO9781139047418.
5. Barringer, 292–96.
6. Nancy H. Ramage and Andrew Ramage, *Roman Art: Romulus to Constantine*, 5th ed., (Upper Saddle River, NJ: Pearson Prentice Hall, 2009), 235–39, http://www.worldcat.org/oclc/465462707.
7. Lancaster, *Concrete Vaulted Construction*, 169.
8. Juvenal, *The Sixteen Satires*, Peter Green, trans., (London: Penguin Books, 1998), 78, http://www.worldcat.org/oclc/961141620.
9. Ramage and Ramage, *Roman Art*, 170–74.
10. Adam, *Roman Building: Materials and Techniques*, 241.
11. Jo-Ann Shelton, *As the Romans Did: A Source Book in Roman Social History* (New York: Oxford Univ. Press, 1988), 308, http://www.worldcat.org/oclc/468344608.
12. Rowland, *Vitruvius: Ten Books on Architecture*, 37.

13. C.J. Brandon, R.L. Hohlfelder, and M.D. Jackson, *Building for Eternity: the History and Technology of Roman Concrete Engineering in the Sea*, ed. John Peter Oleson (Oxford: Oxbow Books, 2014), http://www.worldcat.org/oclc/886881715.
14. Sabbie A. Miller and Frances C. Moore, "Climate and health damages from global concrete production," *Nature Climate Change* 10 (2020): 439-43, https://doi.org/10.1038/s41558-020-0733-0.; "Cement Statistics and Information," U.S. Geological Survey, National Minerals Information Center, https://www.usgs.gov/centers/nmic/cement-statistics-and-information.
15. Rowland, *Vitruvius: Ten Books on Architecture*, 37–38.
16. Vince Beiser, "The Sand and the Fury," *Wired*, June 2015, https://web.archive.org/web/20201108131154/https://www.wired.co.uk/article/the-sand-and-the-fury.
17. Jean-Louis Cohen and G. Martin Moeller, "Introduction," in *Liquid Stone: New Architecture in Concrete*, eds. Jean-Louis Cohen and G. Martin Moeller (New York: Princeton Architectural Press, 2006), 6, https://issuu.com/papress/docs/liquid-stone.
18. Franz-Josef Ulm, "What's the Matter with Concrete," in *Liquid Stone: New Architecture in Concrete*, eds. Jean-Louis Cohen and G. Martin Moeller (New York: Princeton Architectural Press, 2006), 218–42, describes these new techniques.
19. Ulm, 243.
20. Ben Guarino, "Ancient Romans made world's 'most durable' concrete. We might use it to stop rising seas," *Washington Post*, July 4, 2017, https://web.archive.org/web/20201208033810/https://www.washingtonpost.com/news/speaking-of-science/wp/2017/07/04/ancient-romans-made-worlds-most-durable-concrete-we-might-use-it-to-stop-rising-seas/.

Copper and Bronze: The Far-Reaching Consequences of Metallurgy

FLORIN CURTA

"Bronze is the mirror of the form." —Aeschylus, frg. 384

Abstract

Historians and archaeologists have long viewed the discovery of metals and the invention of metallurgy as a revolutionary step in the history of humanity. But metallurgy was more than a technical revolution; its invention in the Bronze Age was primarily a social revolution. This chapter introduces the technological innovations associated with the manipulations of metal by smelting and casting, and the economic and social problems that came with the development of this early metallurgy. At the same time, the chapter highlights the role of trade and its connection to the rise of metallurgical, **proto-industrial centers** across Europe and the Middle East. The ensuing social and political complexities, disparities, and military conflicts are direct results of that connection and of the competition for resources inherent for metallurgy. Finally, the chapter points to the rise of a class of material specialists in society, forerunners to our contemporary engineers.

Introduction

We can make tools, jewelry, toys, kitchenware, furniture, and almost any other item from metal. Useful though metals are, they are sometimes less than perfect for the jobs we need them to do. That is why most of the "metals" we use are not actually metals at all but **alloys**—metals combined with other substances to make them stronger, harder, lighter, or better in some other ways. For example, copper is good in some ways, but bronze is far better. Thus, the exploration of copper in this chapter is a way to introduce not only the discovery of metals in human prehistory, but also the invention of metallurgy, the manipulation and alloying of those metals to meet human needs. Understanding the social and political implications of introducing these new materials—alloys—is crucial for the engineers of the future, who will find ways of using and improving existing materials, and will create new ones.

The Revolutionary Role of Metals and Alloys

The revolutionary aspects of discovering the properties of metals and, particularly, alloys, has long been recognized by scholars seeking to explain the social changes that led to our modern civilizations. As introduced by Gillespie in her earlier chapter on Clay, the idea that even the (pre) history of humanity could be divided into "ages" on the basis of the materials out of which humans made their tools and weapons is relatively new. It originated with Danish archaeologist Christian Jürgensen Thomsen (1788–1865; Figure 5.1).[1]

Figure 5.1 Portrait of Christian Jürgensen Thomsen, the "inventor" of the Three-Age system, in which bronze and iron represent the "Metal Ages," the most advanced stages in the development of human technology. [J.V. Gentner (1849), National Museum of Denmark. Wikimedia Commons.]

He studied Greek and Latin in Paris, and was very fond of the Roman poet Lucretius (99–55 BCE), the author of a philosophical poem entitled *De rerum natura* ("On the Nature of Things"). In that poem, Lucretius describes how the first tools humans used were "hands, and nails, and teeth, and stones and branches torn from trees," before they discovered bronze and iron. With bronze, men "tilled the soil" and "roused the waves of war," before "the sword of iron came forth," and, as they despised "bronze sickle's curving blade," they began to "cleave the earth" with iron.[2] Inspired by Lucretius, Thomsen invented the so-called Three-Age System still used in prehistoric archaeology. That we still refer to one of those chronological divisions as Bronze Age underscores the significance of metallurgy for the history of humanity.

It is easy to overstate the importance of discovering metals and alloys. However, the invention and practice of metallurgy were not the most important developments, as Lucretius and Thomsen thought. On one hand, metallurgy remained in some parts of the world a secondary activity without any substantial cultural and social impact (see below). On the other hand, the same techniques were used both in societies with a low level of organization and in complex societies such as those in Mesopotamia and Egypt.

Metallurgy and Social Innovation

Metallurgy consists of a series of complicated operations, from finding and extracting the metal ore to smelting and processing it with heat and specialized tools. Using metals considerably shortened the time necessary for the production of tools or weapons (previously made through a complicated flintknapping process as discussed in Sassaman's chapter on ceramics), and allowed for the mass production of identical, or at least similar, artifacts. A key component of metallurgy is mastering a great number of physical and chemical reactions and processes, as well as the precise sequence of operation and its timing, the **chaîne opératoire** also mentioned in a different context in Sassaman's chapter. That is why the first metals used for metallurgy were those that could be extracted and processed without too much of an intellectual and technological effort. The Bronze Age came *before* the Iron Age, as Thomsen put it, because iron metallurgy is much more complicated and involves a lot more knowledge and advanced technological processing than bronze metallurgy. In a very fundamental sense it was easier to smelt copper and bronze from copper ore than it was to smelt iron from iron ore.

In certain parts of the world, for various reasons early metallurgical knowledge was lost at some moment in time and the technology for producing bronze had to be reintroduced at a later time. People with metalworking knowledge must have therefore moved around for metallurgy to spread. One way for that to happen involved specialists moving into new territories, thus gradually spreading their knowledge. Another was to have people learning the trade in one place, under the direction of specialists, and then moving to new territories or returning to their own communities with the skills to produce metal objects. Either way, the demand for copper preceded the actual movement of specialists and/or their apprentices. However, both in the Far East (northeastern Thailand and China) and in the New World (Mexico and Peru), copper and bronze metallurgy was invented independently, without any contacts with the centers in the Near East and the Mediterranean region.

In short, metallurgy implies a long process of learning and professional specialization. Unlike pottery, for example, metallurgy cannot be done "on the side" of other external work or domestic activities, at least not when the goal is to produce a large number of similar artifacts. In other words, someone involved, for example, in agriculture on a regular basis cannot do some metallurgy during his or her spare time. The learning process is long and complicated, and that requires time and dedication. That is the main reason that metallurgy cannot spread as a diffusion of ideas from person to person. Rather, metalworking needs to be taught by a specialist to another person. Metallurgy implies not only the existence of specialists, but also a complex process of learning, which sets it apart from other technologies, the spread of which was simply based on diffusion. Put simply, while the *products* made through metallurgy impacted societies by addressing immediate needs, the *process* of metallurgy also impacted society by creating a new class of professional specialists in the social order.

Copper and Smelting

Copper was fashioned into artifacts as early as the 7th millennium BCE. Some of the earliest artifacts—tubular beads and other dress accessories—are from Çatalhöyük in Turkey (ca. 6500 BCE), the archaeological site most prominently featured in the previous chapter on clay. There are good reasons to believe that the copper in those artifacts was extracted from ore rather than created from **native copper**, copper in its pure metallic state. This extraction process is called smelting. A copper axe was found with the mummified body of a man who died in the Austrian Alps ca. 3300–3200 BCE. The axe was smelted from copper ore. "Ötzi," as this extraordinary mummy came to be known, was a hunter, but possibly also a shaman, in which case the axe was not necessarily a tool, but an object used in rituals.[3] (Figure 5.2)

The earliest use of metal, therefore, had no economic role, as most metal artifacts were either dress accessories or objects of ritual use.

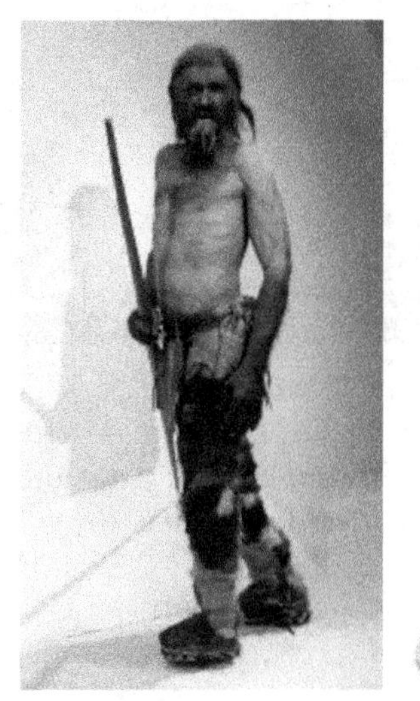

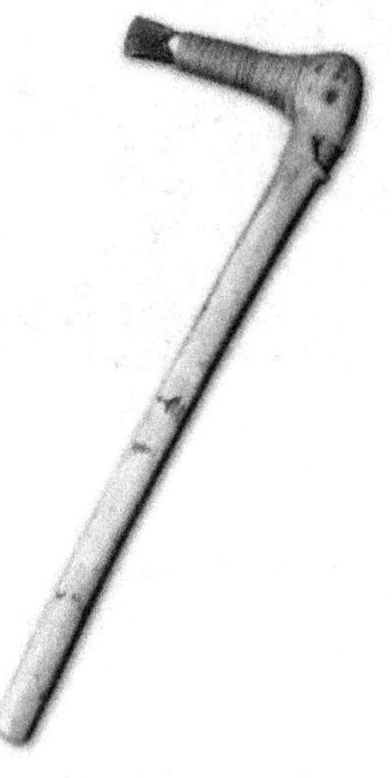

Figure 5.2 The reconstructed copper axe found with the mummified body of Ötzi, the Neolithic hunter, in the Tyrolian Alps, near Innsbruck (Austria). The axe is made of native copper. [Wikimedia Commons.]

Native copper is still available in regions of the world such as Australia, France, China, Namibia, and Iran, where it appears in the form of distorted masses or extremely distorted crystals. Native copper does not have impurities, so it can be shaped by hammering. The copper artifacts from Çatalhöyük were hammered, possibly with stone tools. Hammering can induce dislocations or imperfections in the copper, making it both harder but also more brittle (a process known as work hardening). Applying heat removes some of these dislocations and renders copper less brittle, an operation known as **annealing.** These processes initially served for the fashioning of durable cutting edges (e.g., for axes or knives). Such artifacts were sometimes collected and deposited in the earliest *hoards* known to history, but none of them show any wear and tear, a clear indication that their function was not utilitarian, but ritual. In other words, these early metal artifacts were produced for storing a materials (copper or bronze), which due to the challenges of smelting were rare and therefore precious. Why would metal be regarded as rare and precious? The first worked copper was the native variety that was dug out of the earth's surface (Figure 5.3). Because native copper is not commonly found copper in general was perceived to be rare.

Figure 5.3 Native copper from Ray mine (Arizona). Copper sometimes appears as isometric cubic and octahedral crystals, but more often as irregular masses and fracture fillings. The specimen in the picture is only 5.25 cm long. [Photo by Rob Lavinsky, iRocks.com (2010), shared under a CC-BY-SA 3.0 Unported License. Wikimedia Commons.]

The introduction of impurities in the copper can increase the strength and hardness of the metal. The mixture is called an alloy. One of the most frequent impurities is arsenic, which is present in such ores as arsenopyrite, enargite, and especially tennantite. The addition of arsenic creates a copper alloy, which is called bronze. Tennantite seems to have been a common starting material for obtaining arsenical bronze. During the bronze formation process the arsenic mixed with the reduced copper to create the alloy of bronze, and thus it was possible to obtain copper mixed with impurities without very complicated technologies. Indeed, some of the **halberds** (particular kinds of weapon combining a spear with a battle-axe) that archaeologists discovered in England, Ireland, and China are made of copper mixed with large amounts of arsenic, and have rivets of pure (and therefore more malleable) metal (Figure 5.4).

When the primary copper ores that contain arsenic are heated in a process sometimes called roasting, they turn into copper arsenate (olivenite). Further reduction of the copper arsenate by heating with charcoal yields a copper-arsenic alloy (bronze). During the heating process, however, a very large quantity (over 50 percent) of arsenic is lost as As_2O_3. Roasting and alloy formation may well be a relatively simple technology, but the production of copper-arsenic alloys was challenging due to the difficulty of controlling the arsenic content and the release of arsenic gas, which is highly toxic to the craftsman.

Figure 5.4 Chinese bronze halberd with bird-shaped decoration on top from between the 4th and 3rd centuries BCE (Eastern Zhou Period). [Wikimedia Commons.]

Key Concept: Hoard

Collecting valuables either for display or for reuse at a later time seems like a permanent dimension of human history. In reality, this social phenomenon starts in earnest in the Bronze Age and is directly connected to metallurgy. The earliest such collections contained precious but also practical items (some of them deliberately broken in advance), jewelry, and weapons. The logic behind the Bronze-Age deposits of collections of items has nothing to do with economic security. Most hoards were found in or next to bodies of water and were probably meant to be irretrievable. In other words, such collections are the only surviving parts of complex rituals, which may have involved a conceptual association between valuable metal objects and the divine.

Copper ores are typically of two kinds: carbonates (malachite) and sulfide ores (peacock ore or fahlerz). The distribution of both kinds on the surface of the planet is very uneven. The earliest evidence of copper ore mining was to source copper silicates and malachite (copper carbonate). In order to extract copper from malachite, one needs to separate the metal from carbon and oxygen and other impurities. This was possible only through **smelting**, a new technology for processing copper ores that gained enormous importance in early metallurgy.

A form of extractive metallurgy, smelting is based on the idea of bringing the ore to a temperature sufficiently high for melting the metal. A reducing agent decomposes the ore, thus separating the other elements as gases or slag and leaving the metal base behind. The reducing agent most commonly used in the past was charcoal, which produced CO upon heating in a reducing environment. The reducing environment consisted of an air-starved furnace, in which the incomplete combustion of carbon created CO, which subsequently took the excess oxygen from the metal ore and left the metal behind. Copper smelting involves high temperatures and a reducing environment, two conditions that characterize pottery kilns. It is therefore possible that the process was discovered during experiments with firing pottery. In other words, while playing with "cooking" copper ore, early potters may have discovered smelting, and in the process became the earliest metallurgists.

Activity: Watch a video

Smelting can be done either in a crucible or in a furnace—a single or multiple bowl-like feature over which a clay superstructure was built to contain the ore and the charcoal. This video demonstrates the process of copper smelting in a pit (https://www.youtube.com/watch?v=8uHc4Hirexc).

What were the critical ingredients used in the smelting and what facilities were necessary?

Smelting in furnace first appeared in the Near East, specifically in the lands now within Israel and the Sinai Peninsula (Egypt) during the 5th millennium BCE. Probably the most famous of the earliest smelting sites is Timna, where the earliest evidence of smelting in furnace has been radiocarbon-dated to 4460–4240 BCE.[4] Smelting at Timna released impurities either as gas (CO_2) or as slag (primarily silicates), which was tapped off, while the copper sank to the bottom of the furnace, where it was collected in the form of plane-convex ingots. In the Mediterranean region, however, pure copper was formed into **ingots** of a special form—the so-called "oxhide ingots"—probably in order to be transported at long distance (Figure 5.5).

Figure 5.5 Oxhide ingot from Zakros (Crete), now in the Heraklion Archaeological Museum (Greece). How would you explain the shape of the below ingot? Can you think of any practical reasons for the long sides being curved? [Photo by user Chris 73, (2005), shared under a CC-BY-SA 3.0 Unported License. Wikimedia Commons.]

The earliest such ingots have been found in Hazor (northern Israel) and dated between the 17th and the 16th centuries BCE. Lead isotope analysis revealed that the copper in those ingots actually came from Cyprus, which implies regular and quite intensive commercial contacts between the Mediterranean island and northern Israel.[5] However, oxhide ingots appear as far to the north and to the northwest as Bulgaria and Germany, and they also appear in Egyptian wall paintings.[6] The trade connections made possible by the need to procure good copper for metallurgy seems therefore to have extended very far in a relatively short period of time.

Metallurgy, Alloys, and Trade

Because pure copper is soft and malleable, the idea of adding other minerals may have come from the production of copper-arsenic alloys. The addition of other metals is useful, because it reduces the temperature at which copper melts, while at the same time increasing the hardness of the finished metal. Copper melts at 1,085 degrees Celsius, which implies the use of an enclosed furnace and of forced draught or the introduction of overpressure of air. By adding only 10 percent tin to the alloy (bronze), its melting temperature drops to 1,000 degrees Celsius, which still requires a furnace, but not forced draught. On the Brinell scale of hardness, the value of pure copper varies between 35 (for cast copper) and 110 (for copper altered by means of cold working) (in comparison glass has a hardness of 1550) . The addition of 10 percent tin increases the hardness to a value ranging from 70 (cast) to 230 (when the copper is fully cold-worked). It is perhaps unsurprising, then, that beginning with the 16th century BCE, alloys produced in Transcaucasia (between northern Turkey and Russia) mixed tin with copper, as demonstrated by the metallographic analysis of artifacts found in Stepanakart (Azerbaijan), Sengavit (Armenia), and Mekegni (Daghestan). The metallographic analysis showed that the particular choice of ingredients for the alloy was not just a matter of available materials, but also a component of careful planning, since specific artifacts required specific alloys.

Tin for creating bronze came from tin ore (tinstone or cassiterite), which has an even more uneven distribution on planet Earth. Tin for prehistoric bronzes came from Sardinia, Brittany (France), Cornwall (England), Iran, or Bohemia. There is no tin in the Near East or the Eastern Mediterranean, despite the fact that some of the earliest centers of metallurgy were located there. It's therefore no surprise that by 800 BCE, tin appears in lists of commodities as a precious metal, along with such luxuries as gold, silver, iron, elephant hides, ivory, and purple-dyed cloth.[7] The rarity of tin, as well its importance for some of the hardest and most durable copper-alloys, explains why the development of metallurgy encouraged the development of long-distance exchanges and of trade.

Long-distance trade already existed in the Neolithic period, to circulate lithic materials (such as obsidian, mentioned in Chapter 3) or amber, for example. However, both the density and the intensity of commercial exchanges increased considerably during the Bronze Age, which led to the establishment of "fixed routes" along which goods were moved from northern to southern Europe, and from there to the Near East (western Asia, Turkey and Eqypt) and beyond. The exchange of these material goods encouraged the exchange of ideas, technologies, and ornamental patterns. Bronze-Age human communities, at least in the Old World (Mediterranean region to China), were much more interconnected than ever before. Moreover, the exceptional position of some communities, either close to raw materials or strategically positioned at the crossroads of important trade routes, led to unprecedented levels of economic prosperity, as well as aggression. For example, with no tin resources in the

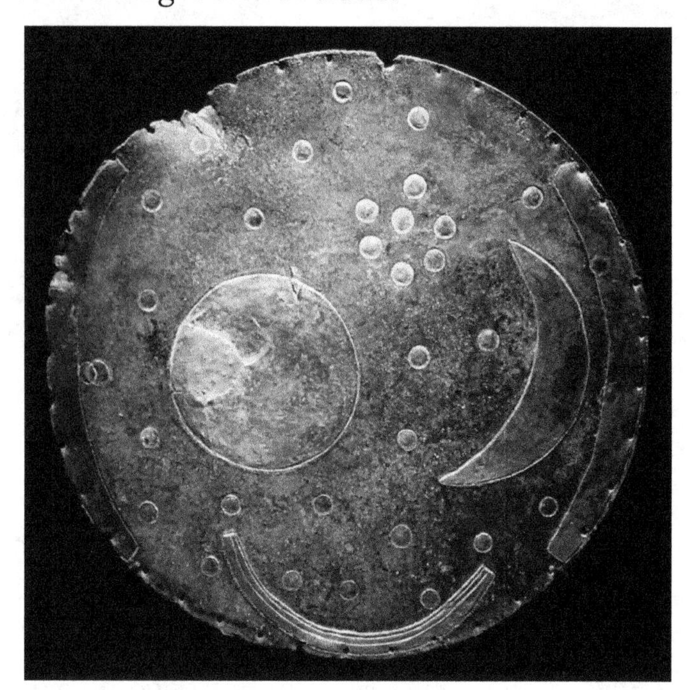

Figure 5.6 The Nebra Sky Disk, a bronze disc of about 30 cm in diameter with representations of the moon, the sun, and the stars. The disc was found in 1999 in Saxony Anhalt, Germany, and is attributed to the Únětice culture (Early Bronze Age, ca. 1600 BCE). The analysis of trace elements by x-ray fluorescence has revealed that the copper in the alloy originated in Austria and the gold in Transylvania. The tin may well have come from Cornwall. The Nebra Sky Disk is therefore a unique illustration of the astronomical knowledge of the prehistoric inhabitants of Central Europe, as well as an excellent example of the connections between bronze metallurgy and trade. [Photo by user Dbachmann, (2006), shared under a CC-BY-SA 3.0 Unported License. Wikimedia Commons.]

vicinity, the archaeological sites attributed to the Únětice culture in Central Europe (2300–1600 BCE) reveal an abundance of bronze artifacts, some of which are ingots, an indication that bronze moved up and down along the trade routes from the Baltic to the Aegean Seas in both raw and manufactured form. Moreover, Únětice sites have also produced evidence of contacts with the British Isles, the main source of the tin that went into the alloy produced on sites in Europe and the Near East (Figure 5.6).

While copper and tin moved in one direction, other goods came from the opposite direction (Figure 5.7). For example, originating from the southern coast of the Baltic Sea, amber is found on many sites in the Near East and Transcaucasia in the form of beads, pendants, or even cups.[8] Many gold artifacts found on Mycenaean sites in Greece are made of metal from Transylvania, while the silver axes from hoards discovered in Romania have a decoration most typical for Mycenaean weapons.[9]

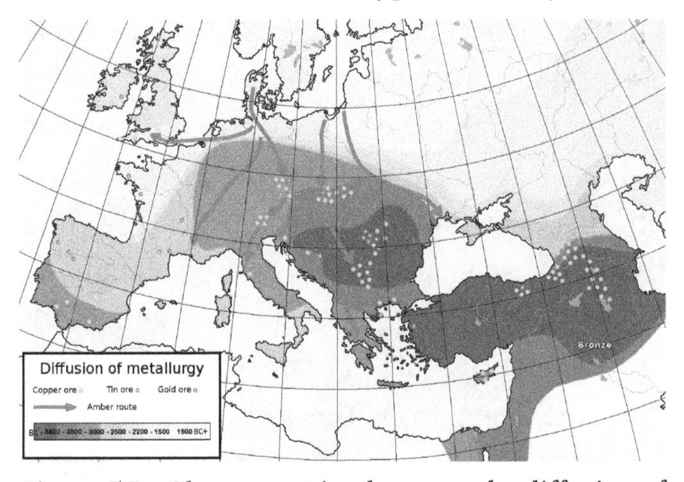

Figure 5.7 The connection between the diffusion of the bronze metallurgy and the development of trade routes is evidence on this map showing the diffusion of metallurgy from 3800 to 1500 BCE. [Wikimedia Commons.]

In the Mediterranean region, long-distance trade is also documented by means of copper and tin ingots from some of the earliest shipwrecks that underwater archaeologists have discovered. A ship went down before 1300 BCE at Uluburun, off the south coast of Turkey, together with a cargo of copper and tin, as well as other rare goods, such as cedar and ebony wood, terebinth resin, elephant tusks, tortoise shells, and ostrich eggshells. There were no fewer than 354 oxhide ingots on the ship, some of which were marked by incision, most likely upon receipt or export, and probably as a warranty of their quality for export.[10] To take another example, the lead isotope analysis of oxhide copper ingots found in Sardinia—an island off the coast of Rome, Italy, with clear evidence of copper smelting in the Bronze Age—showed that they had originated in Cyprus (over 2,000 kilometers away).[11] The invention of metallurgy thus put a high trade premium on metals and spurred the development of long-distance trade in response to a high demand in the emerging centers of metallurgical production.

Metallurgy and Social-political Complexity

The movement of metals in the form of ingots along these long-distance trade routes led to the rise of metallurgical centers in regions otherwise devoid of any local resources for alloying metal. Some of the most intriguing sites with abundant evidence of bronze metallurgy appear in areas with no copper or tin ores. For example, Hacınebi, in southeastern Turkey, was a proto-industrial center specializing in the purification and casting of copper coming from the area farther to the north. The end products were then exported to the southern part of the Near East to be further processed there into local bronzes on the basis of imported tin. The excavations carried out in Hacınebi provided evidence for every aspect and each step of the smelting process, including slags, fragments of slag-accreted crucibles, clay molds, and smelting furnaces. Trade contributed to the success of Hacinebi. But also there was a wealth of knowledge there as well. There is even a fragment of tuyere (a tube through which air is blown into a furnace), the shape of which indicates that it was used at the end of a reed blowpipe, and not for actual bellows. In order for a temperature of 2,192 degrees Fahrenheit to be maintained inside a furnace with an internal diameter of 25 centimeters, no fewer than three adults had to blow continuously into pipes such as those that may have been used for the Hacınebi tuyere. In other words, smelting was a labor-intensive operation and clearly required craft specialization.[12]

In fact, the archaeological evidence suggests that smelting was developed far beyond cottage production and involved the existence of specialized workshops. Scanning electron microscopy was used to examine a slag-accreted crucible from Hacınebi and it was discovered that sulfide ores were utilized for smelting. The copper ores in question came from at least 200 km (124 miles) to the northwest from the site. Metallurgy at Hacınebi required organization and planning, special trade deals with the northern neighbors to ensure a constant supply of ore, and special trade deals with various neighbors to the south, who purchased the copper made in Hacinebi.

Moreover, sites similar to Hacınebi were often fortified, and probably inhabited primarily by smelters who were brought there from somewhere else. This strongly suggests that proto-industrial centers such as Hacınebi could not have existed without some form of political organization guaranteeing the stability of the trade routes and the safety and security of the specialists residing within their walls. Elsewhere, fortified settlements seem to have operated as power centers. For example, Early Bronze Age Lerna (in Greece) had a double ring of defense walls with gates and towers. Inside the fortification, there was a palace or administrative center in a central building that archaeologists called the "House of Tiles" (Figure 5.8). Many proto-industrial centers were also power centers. The invention of metallurgy triggered a whole set of transformations in society, some of them with far-reaching consequences for such things as labor division and the rise of early states.

Geographic factors have long influenced political relationships. There were significant changes in the geopolitics of the Bronze Age. As discussed earlier in this chapter, our knowledge of Bronze-Age tools and weapons is primarily based on deposits (hoards, or caches of objects buried for safe-keeping). Some of the largest hoards have been found in the central European regions of modern-day Hungary and Transylvania (western and central Romania). Although both regions lack tin, throughout the Bronze Age they witnessed the rise of complex societies clearly geared towards war and conquest. By developing contacts with other societies located at a long distance (for example, Mycenae, in Greece), Bronze-Age communities in central Europe were able to procure the raw materials necessary for bronze metallurgy. They also borrowed from their trade partners techniques for the metallurgy of gold and silver, of which they had more abundant resources. As a matter of fact, the development of metallurgy in the Bronze Age involved not only copper, but also gold and silver, as well as lead as will be discussed in the following chapter.

Figure 5.8 Lerna, stairs to the upper floor in the House of Tiles, which probably served as an administrative center inside the fortified site built there in the Early Bronze Age. [Photo by Heinz Schmitz (2006), shared under a CC-BY-SA 2.5 Generic License. Wikimedia Commons.]

With a relatively low melting point at 621 degrees Fahrenheit, lead was one of the easiest metals to process in ancient times. It was first employed for making rivets to be used for repairing broken containers such as found at Phylakopi on the island of Melos and Chalandriani on the island of Syros, both in the Aegean Sea. Lead ingots discovered on both sites indicate a local production of lead, as well as trade with lead in the same form and probably along the same routes as those used for copper and tin. In addition to utilitarian functions, metallurgy served the needs for most sophisticated representation of social status and political power. Lead, instead of tin, was in fact used as a substitution alloy in the production of shaft-hole axes such as found in Belgium and Scandinavia. By contrast, the earliest artifacts made of gold—ring- or disc-shaped pendants—appear in the the 5th millennium BCE on sites attributed to the Gumelniţa and Tiszapolgár cultures of Southeastern and East Central Europe, respectively (Figure 5.9). The jewelry and decorations from the "Royal Tombs" at Alaca Hüyük (Turkey) are among the earliest silver artifacts known and have been dated to the 3rd millennium BCE.

Figure 5.9 A high-status male burial from Varna (Bulgaria), dated to the 5th millennium BCE. The grave goods include both bronze (axes, but also spear heads), and gold artifacts (bracelets, beads, a scepter and a pectoral disc). Note the abundance of semi-spherical gold mounts that likely adorned the shroud that covered the body. [Photo by user Yelkrokoyade (2007), shared under a CC-BY-SA 3.0 Unported License. Wikimedia Commons.]

The Use of Metallurgy: Bronze Artifacts

Unlike gold and silver, bronze was initially used for the production of daggers, axes, and swords. Throughout the Bronze Age, tools continued to be made out of stone. Only later did iron metallurgy put an end to the use of stone as raw material for tools. So what was copper used for at origin?

The first implements made of copper were daggers, probably for ritual and not practical use. Such daggers have been found in Beycesultan and Alaca Hüyük, both in Turkey. The earliest axes were flat or adze-like (e.g., palstave), much like the axe found with Ötzi's mummy. Those were not very efficient tools, and some may have had only a special, probably ritual function. By contrast, the earliest functional axes were shaft-hole specimens discovered on sites of the Tiszapolgár and Gumelniţa cultures. Saws (first documented archaeologically at Los Millares, in Spain) and sickles appeared later, around 3000 BCE and 1500 BCE, respectively. Except for harvesting tools such as sickles, there were no agricultural implements; the plowshare is an Iron Age, not a Bronze Age, invention.

The most impressive artifacts of the Bronze Age are the weapons. The dagger appears to be a Near Eastern invention, and the sword a European invention. Bronze Age swords were made by casting, after which the edges were hammered. This type of weapon originated in the region around, and especially north of, the Black Sea, probably as a further development of the dagger. The technology to produce blades is first documented archaeologically in the Aegean peninsula, where both copper-tin and copper-arsenic alloys were used to produce swords around 1700 BCE. They were of course longer than daggers, with blades in excess of 100 centimeters (over 3 feet).

A great diversity of swords existed in the Middle Bronze Age (1500–1400 BCE, some of which originated in the Mediterranean (Mycenaean Greece) and others in the northwest (Ireland). One of the most important weapon types of the Bronze Age, and the longest-lasting sword of prehistory, is the so-called Naue II type (named after a German archaeologist who first described such swords in the late 19th century). Such swords produced in Europe were highly valued in the Near East, but were quickly replaced after 1200 BCE with iron blades.

While iron blades were also produced in Europe, it was in China that bronze swords remained in use the longest, the latest being produced during the Han dynasty (3rd century BCE–3rd century CE). An equally European origin may be attributed to the battle-axe, which makes its appearance around 1500 BCE, followed after two centuries or so by helmets, shields, and armor (Figure 5.10).[13] Thus it's clear that the desire for ever-better weapons drove much of the innovation in metallurgy.

Figure 5.10 Bronze-Age helmets from Viksø, Denmark, now in the National Museum in Copenhagen. Each helmet is made up of two halves joined by riveting. The horns are riveted by means of fixed circular fixings. Despite popular misconceptions, those were not helmets used in actual fighting (and definitely not by Vikings, who came only 2,500 years later!), but most likely ceremonial helmets for various religious rituals. [Photo by Simon Burchell (2011), shared under a CC-BY-SA 3.0 Unported License. Wikimedia Commons.]

Mold and "Lost-Wax" Casting

The ability to manipulate copper was due to a variety of technological and social developments: trade and professionalization as aforementioned, but also technologies of production such as molding and lost wax casting. This is true for other materials as you will see in the case of iron and steel. For example, molds were used extensively for bronze manufacturing. This relatively rapid development of artifact form and complexity would not have been possible without the parallel development of mold technology. Casting could be done in open one-piece molds carved onto the sides of stone blocks (sometimes even into the native rock). Molds composed of two identical halves were made first of stone, then of clay.

Activity: Watch a video

Claudio Cavazzuti, University of Bologna, demonstrates the process of creating a sandstone mold for casting an Early Bronze Age axe (https://www.youtube.com/watch?v=g5npVVVyWYg).

- What are the advantages and disadvantages of using molds for casting?
- What skill set would craftsmen need to possess in order to cast in molds?

For intricate forms, or for producing parts of larger objects, a new technique was invented ca. 3000 BCE—the "**lost wax** casting approach."

Key Concept: "Lost Wax" Casting

The "lost wax" technique allows for the casting of objects with complicated shapes. The details of that shape are initially rendered in wax, after which the negative space created by melting the wax is filled with molten bronze (1,600 degrees Fahrenheit). This technique also allows for the casting of larger objects, including bronze statues, several components of which could be cast in a sequence of "lost wax"-casting events. During the Renaissance (late 15th century CE), the indirect lost-wax technique was developed, which made it possible to make copies of statues. The surface of the statue was divided mentally into different parts, and covered in clay placed over the designated segmented areas, much like a jigsaw puzzle around a 3D object. When the pieces hardened, the statue was removed, and the pieces reassembled and securely bound together. The empty space was filled with molten wax to create what is known as an "intermodel." After the latter was freed from the piece mold, wax rods (sprues) were attached perpendicular to the surface of the intermodel to serve as the vents for the evacuation of air and gasses during the casting process. Another layer of clay was placed over the intermodel and the whole structure was baked to melt the wax. The resulting mold was then filled with molten metal, as in the traditional "lost-wax" technique.

One of the first free-standing bronze statues since Antiquity, Donatello's *David* (1440s), now at the Bargello Museum in Florence, was cast in this manner. Bronze casting, however, is used not only by artists, but also to preserve original works of art. Leonardo da Vinci's horse (known as *Gran Cavallo*), for example, was part of an equestrian statue of the duke of Milan, Ludovico il Moro (1494–1499), but the artist never managed to finish the work. The statue was meant to be the largest equestrian monument in the world. Based on Leonardo's sketches, two full-size bronze casts were produced in 1998, one of which is now in Milan, the other in Grand Rapids. The indirect lost-wax technique led to a proliferation of copies, which in turn prompted the adoption of prohibitive laws. For example, in 1956, a French law limited the number of copies of each Rodin sculpture to twelve—eight to be purchased by anyone, and four to be in the exclusive possession of cultural institutions.

The Social Implications of Metallurgy

The range of forms to be produced by various casting techniques increased enormously throughout the Bronze Age. Perhaps more importantly, the practice of using the same master object for the production of clay molds allowed for the production of sets of identical end products in bronze. Some forging may have followed the casting, in order to produce sharp edges (as in the case of swords and axes, but not always for sickles), thin blades (of daggers), or to bend items to required shape. For the production of such dress accessories as torcs (neck ornaments), bracelets, or composite rings, wiredrawing was practiced by pulling red-hot metal between draw bars, which thinned the bars down. Thin sheets of copper were produced by hammering metal bars onto an anvil. Both wiredrawing and thin sheet hammering were techniques employed primarily in gold and silver metallurgy. Another technique invented during the Bronze Age for the decoration of objects made of thin gold or silver sheet is the so-called *au repoussé*. With this technique, bosses, dots, rosettes and other motifs were produced by pushing the metal sheet into wooden forms. The technological innovations accompanying the invention of metallurgy thus created a vast field of artisanal expertise, and made room for a conceptual distinction between craft and art and between artisan and artist.

Figure 5.11 The hoard of gold artifacts found in Villena (Spain). The hoard includes of 59 objects and weighs 10 kg, of which nine consist of 23.5-carat gold. The hoard dates from the late Bronze or early Iron Age, ca. 1000 BCE. [Photo by Enrique Íñiguez Rodríguez (ca. 2000), shared under a CC-BY-SA 3.0 Unported License. Wikimedia Commons.]

Much evidence for Bronze Age metallurgy comes from the analysis of hoards that were deposited in locations from which they were never retrieved for a wide variety of reasons. There are also hoards of gold artifacts, with a large array of spectacular objects, such as the bracelets, diadems, and gorgets from Villena in southern Spain (Figure 5.11).

Besides being an illustration of the parallel development of bronze and gold metallurgy—often with comparable techniques—the hoards testify to the concern with accumulating (and storing) wealth, which must have been a direct consequence of the social transformations triggered by the introduction of metallurgy (see Chapter 6). One of the most famous golden objects of the European Bronze Age, for example, is the mask found in 1876 in a cylindrical shaft grave inside the fortified settlement at Mycenae in southern Greece (Figure 5.12). The mortuary use of this gold mask is not unique; in fact, many gold and bronze objects (especially weapons) were deposited in graves. The northwest European equivalent of the rich burials in Mycenae is a number of large, circular barrows that appear in the Middle Bronze Age on sites of the Wessex culture in England. Early Bronze-Age barrows in Transcaucasia were often reserved for individuals of high status, buried with a large collection of artifacts, including chariots with wooden wheels.

In the eyes of many, the complicated technological procedures involved in smelting and casting must have turned the full-time specialists in bronze metallurgy into individuals with extraordinary powers. The civilizing hero of Greek mythology, Prometheus, was a metallurgist providing "divine knowledge" to humanity. In ancient Roman religion, Vulcan (the god from whose name the English word "volcano" ultimately derives) was said to have stared for hours at the fire, before discovering that, when making the fire hotter with bellows, certain stones sweated silver or gold. Vulcan later fashioned thrones for all gods of the Roman mythology.

Thor, one of the main gods of the Norse mythology, is known for his hammer (Mjölnir), which was custom-made by the dwarves of Nidavellir, the quintessential blacksmiths of the North. Both in antiquity and in contemporary times traditional societies (such as those in East Africa, for example), smelters enjoy a much more prestigious status than smiths.[14]

Part of the explanation for this may be that while the work of smiths is a fairly clear process of turning molten metal into objects, the work of smelters looks more like magic as it turns "stones" into metal ingots.

Figure 5.12 The gold death-mask from shaft grave V, grave circle A, in Mycenae. The mask was discovered in 1876 by Heinrich Schliemann, who believed the grave to have been that of Agamemnon. However, the mask is most likely from the second half of the 16th century BCE and thus at least 200 years older than the historical character believed to have been the model for the king of Mycenae known from the Homeric poems. [National Museum of Athens; photo by Xuan Che (2010), shared under a CC-BY 2.0 Generic License. Wikimedia Commons.]

Conclusion

Alloys of a metal are generally stronger and harder but less ductile than the base metal. The invention of metallurgy during the Bronze Age, which was primarily geared towards the production of alloys, had several crucial consequences for societies of that time and later.

On a technological level, a series of new skills became at the same time necessary and commonplace. Such skills required long-term learning processes and apprenticeship, which transformed a group of people in society into specialists, and set apart their social position, both in lifetime and in death. A good parallel to that is the way in which nuclear energy is put to use in the modern world. It is simply not sufficient to build a reactor; a number of highly trained specialists are required as well. The modern use of nuclear technology offers another parallel to the early bronze metallurgy. The potential for use in military applications often advances the non-military or civilian applications.

From an economic point of view, even though bronze was not used for the production of tools as much as iron would be during the Iron Age, raw materials (copper, tin, lead in the form of ingots) and finished products (weapons or tools made of bronze) became more abundant. The early history of metallurgy also reveals the connections between technology and the rise and development of trade routes. Not only does knowledge still spread along trade routes, but there are still very good examples of industrial powerhouses developing in regions of the world devoid of resources, much like in the Early Bronze Age. For example, "silicon valley" did not form because there was an abundance of silicon present there. Rather it developed because innovators and engineers who could manipulate silicon were there.

Perhaps more important from the point of view of this course is the fact that the introduction of metallurgy had social implications far beyond setting apart a class of specialists. The new artifact types made possible by metallurgy introduced the possibility of new scales of value and new ways to demonstrate social divisions. Accumulating objects made of bronze, gold, and silver was not just a way to store wealth. It was also a new way to set a group of people apart from the others, not as specialists, but as elites made socially distinct by means of economic privilege. The social disparities created by the adoption of new technologies and their use for the display of power symbols and social status, as illustrated by the archaeology of the Bronze Age, are still a matter of great concern in the 21st century.

Future engineers and their collaborators have much to learn from the lessons of the past. The introduction of bronze—a new technology—called for an unprecedented development of long-distance trade. Mastering the new technology required time- and energy-consuming training of a class of specialists that, for the first time in history, came to play a role in society clearly marked ideologically by their association with magic. Metallurgy also opened new paths for the development of warfare and the symbolic representation of power. Similarly, we should expect any new materials to change the trade patterns around the globe, to create new social categories and inequalities, and to have consequences in fields of human activity that may not have yet been designed.

Discussion Questions

1. During the Bronze Age, objects of bronze, gold, and silver produced through complicated techniques were hoarded as valuables. What counted more for their being regarded as such—the intrinsic value of the metal, or the labor involved in producing them? Explain your answer.

2. Silicon has a much wider distribution on the planet than either copper or tin, for it is the second most abundant element of the Earth's crust. However, there is currently a high demand for silicon primarily because of its use in building materials (such as Portland cement), the semiconductor industry, and solar panels. Can you think of cities similar to Hacınebi, the Únětice culture, or Uluburun that were created from the worldwide trade with silicon, and its associated industries?
3. Are there any groups of specialists in 21st-century societies around the globe that are remotely similar to the smelters of the Bronze Age? If so, who and in what ways?

Key Terms

alloy
annealing
ingot
metallurgy
native copper
proto-industrial center
smelting

Further Reading

Hanks, Bryan K. and Katheryn M. Linduff, eds., *Social Complexities in Prehistoric Eurasia. Monuments, Metals, and Mobility*. Cambridge/New York: Cambridge Univ. Press, 2009. https://doi.org/10.1017/CBO9780511605376.

Mei, Jianjun, and Thilo Rehren, eds., *Metallurgy and Civilisation. Eurasia and Beyond. Proceedings of the 6th International Conference on the Beginnings of the Use of Metals and Alloys* (BUMA VI). London: Archetype, 2009. http://www.worldcat.org/oclc/928997914.

Pare, C.F.E., ed., *Metals Make the World Go Round. The Supply and Circulation of Metals in Bronze-Age Europe. Proceedings of a Conference Held at the University of Birmingham in June* 1997. Oxford: Oxbow, 2000. http://www.worldcat.org/oclc/45870233.

Weeks, Lloyd R. *Early Metallurgy of the Persian Gulf. Technology, Trade, and the Bronze Age World*. Leiden/Boston: Brill, 2004. http://www.worldcat.org/oclc/1025038005.

Yener, K. Aslihan. *The Domestication of Metals. The Rise of Complex Metal Industries in Anatolia.* Leiden/Boston: Brill, 2000. http://www.worldcat.org/oclc/43569376.

Author Biography

Florin Curta, Ph.D. in History (1998), Western Michigan University, is Professor of Medieval History and Archaeology at the University of Florida. He has published five monographs, over 40 chapters in collections of studies, and more than 100 articles. He is also editor of six collections of studies.

Notes

1. Bo Gräslund, "The background to C. J. Thomsen's Three Age System," in *Towards a History of Archaeology*, ed. Glyn Daniel (London: Thames & Hudson, 1981), 45–50, http://www.worldcat.org/oclc/7757964.
2. Lucretius, *On the Nature of the Universe*, trans. Ronald Melville (Oxford: Oxford Univ. Press, 1997), 173, http://www.worldcat.org/oclc/858521823.
3. Michel Carrier, "Ötzi: the mummy from the cold," *Arts and Cultures* 5 (2004): 47–59.
4. Beno Rothenberg and Tim C. Shaw, "Chalcolithic and Early Bronze Age IV copper mining and smelting in the Timna Valley (Israel): excavations 1984 and 1990," in *Ancient Mining and Metallurgy in Southeast Europe. International Symposium, Donji Milanovac, May 20–26, 1990*, eds. Petar Petrović and Sladana Đurdekanović (Belgrade/Bor: Archaeological Institute/Museum of Mining and Metallurgy, 1995), 281–94, http://www.worldcat.org/oclc/490516020.
5. Noël H. Gale, "Copper oxhide ingots and lead isotope provenancing," in *Metallurgy: Understanding How, Learning Why. Studies in Honor of James D. Muhly*, eds. Philip P. Betancourt and Susan C. Ferrence (Philadelphia: INSTAP Academic Press, 2011), 213–20, https://doi.org/10.2307/j.ctt3fgvzd.
6. Anthony Harding, "Oxhide ingots in the European north?" *Antiquity* 89 (2015), no. 343, 213–23, https://doi.org/10.15184/aqy.2014.5.
7. James D. Muhly, "Sources of tin and the beginnings of bronze metallurgy," *American Journal of Archaeology* 89, no. 2 (1985): 275–91, https://www.jstor.org/stable/504330.
8. Anna J. Mukherjee et al., "The Qatna lion: scientific confirmation of Baltic amber in late Bronze Age Syria," *Antiquity* 82 (2008), 49–59, https://doi.org/10.1017/S0003598X00096435.
9. Anthony Harding, "Trade and exchange," in *The Oxford Handbook of the European Bronze Age*, eds. Harry Fokkens and Anthony Harding (Oxford: Oxford Univ. Press, 2013), 377, http://www.worldcat.org/oclc/859338044.
10. Andreas Hauptmann, Robert Maddin, and Michael Prange, "On the structure and composition of copper and tin ingots excavated from the shipwreck at Uluburun," *Bulletin of the American Schools of Oriental Research* 328 (2002): 1–30, https://doi.org/10.2307/1357777.
11. Noël H. Gale, "Copper oxhide ingots: their origin and their places in the Bronze Age metal trade in the Mediterranean," in *Bronze Age Trade in the Mediterranean. Papers Presented at the*

Conference Held at Rewley House, Oxford, in December 1989, ed. Noël H. Gale (Jonsered: Paul Åströms förlag, 1991), 197–239, http://www.worldcat.org/oclc/489688081.

12. Hadi Özbal, Annemie Adriaens, and Bryan Earl, "Hacınebi metal production and exchange," *Paléorient* 25, no. 1 (1999): 57–65, https://doi.org/10.3406/paleo.1999.988.
13. Marianne Mödlinger, "European Bronze-Age cuirasses. Aspects of chronology, typology, manufacture, and usage," *Jahrbuch des Römisch-Germanischen Zentralmuseums*, 59 (2012): 1–49, https://doi.org/10.11588/jrgzm.2012.1.15311.
14. Duncan Miller, "Smelter and smith: Iron-Age metal fabrication technology in southern Africa," *Journal of Archaeological Science* 29, no. 10 (2002): 1083–1131, https://doi.org/10.1006/jasc.2001.0758.

Gold and Silver: Precious Metals and Coinage

FLORIN CURTA

"Gold is money; everything else is credit." —John Pierpont Morgan

Abstract

Until the late 20th century, there were very few "technological" applications of precious metals. Despite the occasional use of silver in medicine, or of gold in dentistry and the production of stained glass, the primary use for both metals was the manufacturing of jewelry, sacred vessels (such as liturgical vessels used in the Church), and coins (made with gold and silver because of their high luster, pleasant colors, and tarnish resistance). Complex societies, such as those in ancient Egypt or Mesopotamia, used both gold and silver as money (standard of value and a means of exchange) but not as coins. Coins were invented in three different places at three different times, but the earliest coins were in fact struck not in gold and not in silver, but in a naturally occurring alloy of both, known as electrum. When in the 6th century city-states in Greece began to strike coins in large numbers, the metal they chose was silver, with gold coming into use on a large scale only in the 4th century BCE. Coined money is based on the idea that the quantity and quality of the metal in the coin is guaranteed by some authority, often that of the state. The coin thus circulates at a value higher than that of the metal from which it is made. This chapter discusses how coined money introduced the conceptual distinction between **intrinsic** and **extrinsic value**. After being used for thousands of years for jewelry, gold is now used in electronics, as well as medicine, which has increased its extrinsic value. However, human sentiment also can increase the extrinsic value of gold. A ring may be an heirloom, an object passed from one generation to another within the same family, which means its value might be defined primarily in social terms rather than economic terms. In other words, the extrinsic value attached to it by means of memories and feelings far exceeds its intrinsic value deriving from a certain quantity of gold or silver of which it is made. This chapter explores how the use of silver and gold as a means of exchange, as well as the invention of coined money, have created a distinction between intrinsic and extrinsic value of which modern engineers need to be aware when considering any new applications of material.

Introduction

The distinction between intrinsic and extrinsic value in the case of gold and silver derives from their use for coined money. Coins came into being in three different places at three different times: Asia Minor in the 6th century BCE, India in the 5th century BCE, and China in the late 3rd century BCE. In all three regions, the invention of coined money had a considerable impact upon the development of society. In all three regions, under the illusion that money is an essential part of the human condition, people began to define happiness as the possession of a as large a quantity of coined money as possible. But "money" is not the same thing as "coin," which in turn has to be distinguished from "currency." Money refers to anything that may serve as means of exchange and of storing wealth. Salt bars, for example, were used as money in Ethiopia and Eritrea until well into the 18th century.

Amole salt bars from Ethiopia [MoneyMuseum, Zürich]

This video explains the use of salt bars as money, through a guided tour of the Geldmuseum, or Money Museum, in Frankfurt, Germany (https://www.youtube.com/watch?v=gXxjdGhv54I).

Currency commonly refers to a form of money, the value of which is guaranteed within a given territory at a given time. For example, the dollar is the US currency, while the euro serves the same purpose for some members of the European Union. Coins are both a form of money and a form of currency. A currency such as the dollar exists in the form of both banknotes and coins (pennies, nickels, dimes, and quarters).

Key Concept: How did money come into being?

Some 2,400 years ago, the Greek philosopher Aristotle (384–322 BCE) believed that "when the inhabitants of one country became more dependent on those of another, and they imported what they needed, and exported what they had too much of, money necessarily came into use. For the various necessaries of life are not easily carried about, and hence men agree to employ in their dealings with each other something which was intrinsically useful and easily applicable to the purpose of life, for example, iron, silver, and the like. Of this the value was at first measured simply by size and weight, but in process of time they put a stamp upon it, to save the trouble of weighing and to mark the value" (*Politics* I 9.7–8). In another work, the Greek philosopher has a different explanation: "The builder must get from the shoemaker the product of his labor, and must hand over his own in return. If, first, proportionate equality is established, and then reciprocation takes place, the result we mentioned will follow. If not, there is no equality, and the bargain falls through, since there is no reason why what one produces should not be more valuable than what the other produces, and the products must therefore be equated. . . This is where money comes in; it functions as a kind of mean, since it is a measure of everything, including, therefore, excess and deficiency. It can tell us, for example, how many shoes are equal to a house or some food. Then, as builder is to shoemaker, so must the number of shoes be to a house" (*Nicomachean Ethics* V 5.8–10). One definition is based on the idea that money facilitates trade, the other definition emphasizes money as a standard of value. Money is a substitute for all goods, but also something that is needed for human justice. The second definition implies that money is not just a practical solution to problems of exchange, but also a necessary condition for a certain level of social complexity. Over the last two millennia, philosophers and historians have debated which of the two definitions should be preferred. Both, however, highlight the far-reaching economic and social implications of money.

Wealth was not always stored or measured in money. Even today, a wealthy person is said to be "free from pecuniary anxieties." Few know, however, that the adjective "pecuniary" employed in that phrase derives from the Latin word for wealth or money (*pecunia*). Even fewer know that that Latin word derives from the word *pecu*, which refers to a flock of sheep or, in general, to cattle. How did cattle come to be associated with money? The German economist and historian Bernhard Laum (1884–1974) first noticed that in the *Iliad* and the *Odyssey*, the worth of most items is expressed in cattle, even though cattle did not serve as a means of exchange. An ox in the *Iliad* was not only "wealth," but also a sacred animal, most appropriate for a ritual sacrifice or for a royal gift, but not for market exchanges. Both bride price (a sum of money that in some traditional societies the groom pays to the bride's family) and dowry (money that in traditional societies a bride or a bride's family gives to the groom after the wedding) were paid in cattle. The same is true as a fine for killing, originally a sacrifice meant to appease the victim's soul. According to Laum, the origins of money as a standard of value developed therefore not from trade, but from the sacrificial practices of Classical Antiquity. The idea of money developed not from exchange between humans, but from the status of the sacrificial animal (the ox) as a substitute for the person performing the sacrifice. In other words, an ox replaced a person, but could in turn be replaced by something else, such as cakes, spits, cauldrons, or tripods (Figure 6.1).[1]

Figure 6.1 A sacrificial tripod offered to Apollo at Delphi. Painting on a ceramic, Red-Figure bell-krater from Paestum, ca. 330 BCE. [Wikimedia Commons.]

Laum's theory shifts the emphasis from a pragmatic explanation of money to one that stresses its profound symbolism. People chose certain objects made of certain metals to represent money because they had a distinct idea about what the equivalent should be for all goods, and what the general measure of value should look like. For money to exist, it had to fulfill four major functions. First, it had to serve as a means of exchange, for without money, there is no complex economy. Second, it had to serve as a standard of value, for without it, buildings, shoes, and any other goods could not be compared to each other. Nor could the labor invested by humans in such goods be compared, so the relative, social value of individuals would be difficult to assess. Third, money must serve as a means of storing value. In the absence of any means to refrigerate or salt his catch, a fisherman must sell his fish quickly. The only way for him to "store" the fruits of his labor is therefore to "turn" the fish into money, by means of a fish sale. Incidentally, that is true, albeit on a more symbolic level, for athletes who compete for Olympic gold, as well as for actors and film directors who aspire to win an Oscar—literally a golden statuette. Finally, money has to serve as a means of payment. The fine for killing mentioned above, the bride price, and the dowry are all "payments" that would not be possible without money (Figure 6.2).

All four functions of money have a long history either separately, or in combinations of two or three functions. However, it was the invention of coinage that combined all four functions together for the first time, and radically transformed economy and society. That transformation is ultimately responsible for the economic distinction between the intrinsic and extrinsic (or monetary) value of gold and silver.

Figure 6.2 Money (in the form of banknotes and gold jewelry) in a formal presentation of the bride price at a Thai engagement ceremony. [Photo by user Tainscough (2008), shared under a CC-BY-SA 3.0 Unported License. Wikimedia Commons.]

Money Before Coinage

Societies in the ancient Near East—Egypt, Babylon, Assyria, Phoenicia, and Israel—used standards of value along with precious metals as a means of payment. However, there were no coins in any of those societies, all of which knew the four functions of money, but did not need them combined into a single object. A court deposition document from the reign of Ramses II (13th century BCE) describes how a merchant sold to an Egyptian lady named Erenofre an enslaved Syrian girl at the price of 4 *deben* 1 *kite* (about 373 grams) of silver. In order to pay the price, Erenofre did not actually give the merchant any silver. Instead she made up a collection of clothes and blankets to the value of 2 *deben* 2 1/3 *kite*, and then borrowed from her neighbors a miscellany of objects—bronze vessels, a pot of honey, 10 shirts, 10 *deben* of copper ingots (see Curta, "Copper and Bronze") to make up for the difference.[2] Similarly, in ancient Mesopotamia, people kept accounts in silver. As in Egypt, silver by weight was a standard means of accounting for the value of different goods. Silver was also used as a means of payment in commercial transactions. When about to make a payment, one need only weigh out the silver. If the weight was uneven, an item was chopped up to make the scale balance, as indicated by pieces of cut silver (hack-silver) found in hoards, such as that buried in el-Amarna (Egypt) at some point during the 14th century BCE (Figure 6.3).

Figure 6.3 Hacksilver in the el-Amarna hoard. The Egyptian word for silver (hedj) came to mean something close to "money." [British Museum. Wikimedia Commons.]

Silver therefore served as money, and both kings and temples established the weight standards and published in inscriptions the values of certain commodities in silver, as well as the amounts to be paid for fines, interest, or wages. However, silver had to be imported into the region, for neither Egypt nor Mesopotamia had any easily accessible sources of that metal. Much of the imported silver was consumed, and even hoarded, by kings, aristocrats, and temples through taxes, tribute, and plunder. Because of its hoarding and consumption, silver had a high extrinsic value and a strong symbolic association with royalty, power, and wealth. No distinction existed between silver as money and silver as jewelry or utensils. Circlets of gold, such as those found in Samaria (Israel) and dated to the early 4th millennium BCE, also point to the lack of that conceptual distinction.[3] Indeed, while meant to facilitate trade, such circlets ended up in hoards (and thus withdrawn from circulation and trade) because of the strong association between precious metals, power, and wealth.

Money in Greece

Gold circlets were not used in Greece in the 8th century BCE. To be sure, Greeks had been in contact with Phoenicians, who used silver as a medium of exchange. But instead of silver, Greeks used cauldrons made of bronze as a medium of exchange: the wealthy were those who possessed many cauldrons and tripods. In addition, iron spits, such as those that archaeologists found in burials, may have served as money. In fact, the smallest silver coins of the Classical age were called *oboloi*, which means "spits" and six of them made up a drachma (*drachme*), which means "a handful of spits" (Figure 6.4).

However, unlike silver in Mesopotamia and Egypt, the bronze cauldrons, tripods, and iron spits of ancient Greece never served as general tokens for market exchanges. They were, in fact, utterly inappropriate for that, since they were bulky and heavy. Cauldrons, on the other hand, were still cauldrons, and spits were spits—objects with practical utility. This is in sharp contrast to utensils used in exchanges elsewhere. For example, in China, spades or knives used as money (*bu*) were completely useless—nothing could be cut with any of them. In that case, it was precisely because they could not be used as utensils that those artifacts were meant to serve as money. By contrast, the utilitarian aspect of those artifacts of ancient Greece that served as means of exchange was never lost. So while money existed in ancient Greece before the invention of the coins, none of the objects chosen for that purpose incorporated all four functions of money.

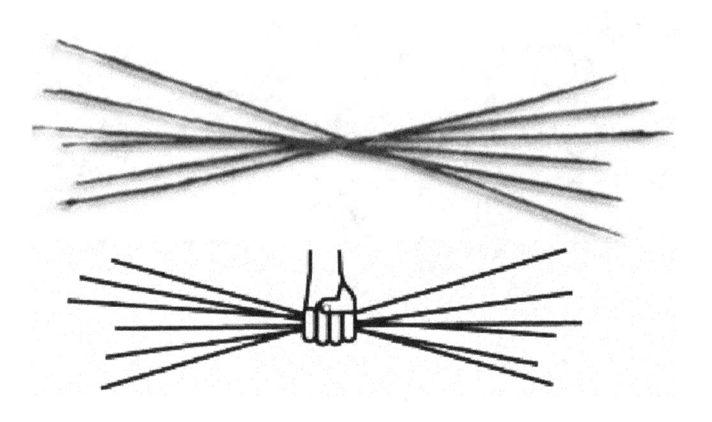

Figure 6.4 Six iron spits discovered in the Temple of Hera in Argos (Greece), 7th–6th century BCE. *A handful of (six) oboloi (spits) formed one drachma. [Wikimedia Commons.]*

The Earliest Coins

Long-distance trade and contact with the Near East must have brought to the Greeks the idea of using precious metal as money. Gold and silver were considered valuable possessions in themselves, as shown in the Homeric poems. In the *Iliad*, Troy is praised for its "wealth of gold," Thetis is "silver-footed," while Athena is "golden-haired." However, silver and gold are conspicuously absent from the archaeological finds from Troy VII, which is believed to correspond chronologically to Homeric Troy.[4] Moreover, gold was not readily available in Greece, as depicted in the legend of Jason and the Argonauts, who had sailed very far from Greece to search for the Golden Fleece.[5] The limited supply of silver and gold may explain why both metals were valuable in the first place. But it may also explain why the earliest coins appear in regions rich in silver and gold. Privileged access to precious metals coincided with the development of political power.

Both the circumstances and the region in which coins were invented are associated with political power. Small bits of metal turned into coins when an impression was hammered or stamped on each one of them (when hammered, such an impression is called "incuse"). It was not the intrinsic value of the metal that mattered, but the extrinsic value of the coin, for which the stamp served as a warranty. The earliest coins were actually made of neither silver nor gold. They were struck in electrum, a naturally occurring alloy of gold and silver that appears in the area of Mount Tmolos (now Bozdağ near Izmir, in western Turkey), which was part of ancient Lydia (Figure 6.5).

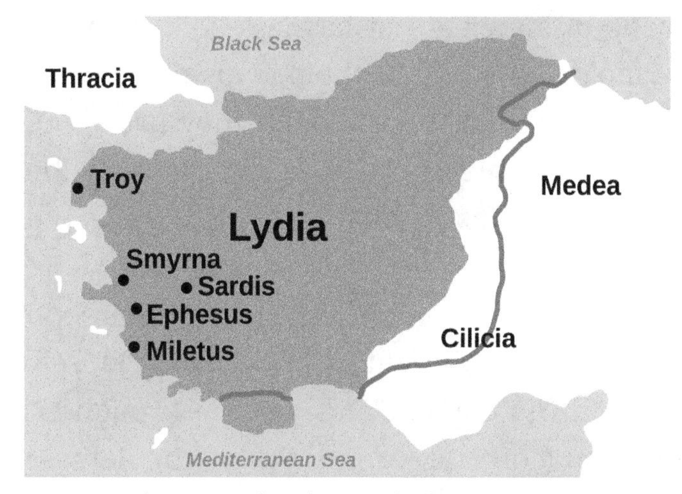

Figure 6.5 Map of Lydia, with the most important Greek cities in existence during the 6th century BCE. [Illustration by user Roke (2013), shared under a CC-BY-SA 3.0 Unported License. Wikimedia Commons.]

The king of Lydia, Croesus (560–547 BCE) was legendary in antiquity. After conquering most of the Greek cities on the coast, the king wanted to show his generosity to his new subjects. At some point in the 7th century BCE, a large flood had destroyed the Temple of Artemis in Ephesus. Croesus paid generously for its reconstruction. The new temple attracted a great number of merchants, kings, and visitors, and many of them honored the goddess by offering her precious goods and jewels. In the early 20th century, excavations inside the temple led to the discovery of a hoard, which included 93 small pieces of electrum and seven silver nuggets. The electrum pieces have a rough oval shape and seem to be aligned to a system of regular weights: the larger coins are 17.2, 16.1, and 14.1 grams in weight, while the smallest ones are only 1/96 of a larger piece. Those were therefore not just metal nuggets. Moreover, three of them have been marked on one side with an incuse square. Other coins display a figurative design, such as a lion, a stag, or a ram. The head of a lion was the symbol of the Lydian royal house (Figure 6.6).

Twenty coins in the hoard have the letters ".WALWE." inscribed on them, which indicates that they were struck in the name of Alyattes, Croesus' father.[6] The lion head and the incuse squares were probably royal markers guaranteeing the quality or fineness of the metal. The small size as well as the standardized weight strongly suggest that the purpose of those coins was to make payments in small amounts, for example to the many mercenaries the Lydian king employed. One of those mercenaries probably made an offering in the Temple of Artemis, and the small coins he had received from the king in exchange for military service turned into a votive donation to the goddess. If so, then the transformation of a payment to the mercenary into a

Figure 6.6 Electrum coins from Ephesus (6th century BCE). One on them was stamped with the image of a lion head, the other has two punches in the form of incuse squares. [Wikimedia Commons.]

"payment" to a goddess suggests that the coins were money, i.e., a means of exchange. In other words, they could have been exchanged for goods on the market. Some scholars have noted that the coins in question are made of an alloy that has less gold than the electrum occurring naturally around Mount Tmolos. The role of the stamps or punches was therefore to make those coins pass for real electrum, even though they had more silver than gold. In other words, the purpose of the punches was to make the metal be recognized and accepted at "face value" rather than by weight, as was the practice in the Near East.

Of course, the coins had some **intrinsic**, material value as bullion. Nevertheless, as soon as the metal had been minted, it gained a different, symbolic value as a coin, only loosely (and, as it were, deceivingly) connected to its metallic content and purity. Throughout the subsequent history of coinage, both the bullion (contents or purity) and the defined weight would repeatedly be altered, as rulers "inflated" the coins and then from time to time had to introduce mint reforms to restore their acceptability. However, not all stamps on the Ephesus coins were royal, and some may have been private. Some of the earliest coins bear inscriptions in Greek, which read "I am the sign of Phanes" or, on smaller pieces, just "Phanes." Some of the designs on the electrum coins may also be found on seals and rings of the same period, which strongly suggests that the coins were struck for private, not state, authorities. Moreover, there were no fewer than eight denominations in the Ephesus hoard, an indication that the coins were instruments for payment, either to the state or on the market. The extrinsic value of the coins, guaranteed by their stamps, was therefore of a greater significance in establishing the value of those objects in relation to any other commodities than the value of the metal of which they were made.

From Silver to Gold Coins

Electrum coins had a short life. Croesus began to strike coins in gold and silver, and by the time his kingdom was conquered by Persians in 546 BCE, coins were already in use in all Greek cities on the western coast of what is now Turkey. From there, they were adopted by cities in Greece as well. By the end of the century, there were more than 100 mints operating in Greece. Most of the coins struck in those mints circulated only locally. However, coins from Thrace and Macedonia (both regions rich in silver) appear in hoards discovered outside Greece, a clear indication that the value of those coins was recognized outside the area of their local circulation. The first coins struck in very large quantities were those of Aegina, an island off the coast of Attica, which has no silver resources. Athens, using the silver from the Laurion mines that opened in the 6th century BCE, first struck **didrachms** before issuing the famous "owl coins," which became the first international currency in the eastern Mediterranean region (Figure 6.7). Those were also the first widely known, head-and-tail coins (i.e., coins with different images for the **obverse** and the **reverse**, respectively), which remained virtually unchanged for 500 years. Within less than a century, the new invention—coined money—spread rapidly throughout the entire Greek-speaking region of the eastern Mediterranean.

The silver coins of Athens also moved to Egypt, the Near East, and the region of the Black Sea—all areas connected to the city by means of trade. As the Greek historian Xenophon put it, "it is sound business to export silver; for where they [the merchants] sell it, they are sure to make a profit on the capital invested."[7] Xenophon's point is very important: silver was exported from Athens as bullion, not as coin. The evidence from hoards found in Egypt shows that instead of being treated as coins, the stamped pieces of metal that reached that country by means of trade were chopped and treated as hack-silver with complete disregard for their extrinsic value guaranteed by the stamps.

Figure 6.7 "Owl coin" (silver ***tetradrachm****) struck in Athens between 480 and 420 BCE. In Greek mythology, the little owl (Athene noctua) was the companion of Athena, the goddess of wisdom and the patron of the city of Athens. [Museum of Fine Arts of Lyon; photo by Marie-Lan Nguyen (2009), shared under a CC BY 2.5 Generic License. Wikimedia Commons.]*

Throughout the Hellenistic period (323–31 BCE), coins tended to have a higher value within their area of issue than the bullion from which they were made. This may explain why coinage in silver spread so quickly in Classical Antiquity. City-states could increase their revenue by controlling the production of silver and the circulation of overvalued pieces of metal struck as coins. That is why the designs chosen for the Greek coins are in fact symbols of the city-state in which they were struck (e.g., the owl as a symbol of Athens) and are often accompanied by inscriptions giving the city name.

Figure 6.8 The ***stater*** *of Eucratides, the largest surviving coin from antiquity. Found in Bukhara (Uzbekistan), the coin weighs 169.2 grams, and is 58 millimeters in diameter. [Cabinet des Médailles; photo by World Imaging, shared under a CC-BY-SA 3.0 Unported License. Wikimedia Commons.]*

City-states had the legal authority to enforce the overvaluation of silver when paid in the form of acceptable coin. Moreover, many of them were highly attractive markets visited by merchants from all over the Mediterranean region. It is those merchants trading in silver or in some other commodities that prevented the state from incurring the great losses involved in accepting overvalued silver coins. Indeed, coinage was produced especially by those city-states that, like Athens or Aegina, had a strong "trade balance" in their favor. That also explains why the coins of those two city-states became the first international coinages of the ancient world. Access to the rich gold mines in Thrace was the main reason for which the king of Macedonia, Philip II (359–336 BCE) and his son, Alexander the Great (336–323 BCE) struck large numbers of gold coins. In addition, through his conquest of the Persian Empire, Alexander gained control of the enormous wealth accumulated over the years by the Persian kings. Alexander's successors (the so-called Diadochi) had also access to enormous resources and wealth, and they struck coins in gold, rather than silver. The largest gold coin of the ancient world was struck in the name of Eucratides (ca. 170–145 BCE), who ruled over a large portion of Central Asia centered upon present-day Afghanistan.

During the Hellenistic period, the portraits of the kings came to replace the old symbols of the city-states. Awarded divine status, the kings of Ptolemaic Egypt and Seleucid Syria issued large gold coins with their portraits on the obverse and the symbols of their kingdoms on the reverse. This established the head-and-tail model of coinage that has persisted throughout the medieval and modern age.

For most of its history, the currency of Rome, both during the Republic and during the Empire, consisted of gold, silver, and bronze. The main reason for this monetary system employing three different metals (trimetallism) is the fact that "good" coins with high intrinsic value (gold, but also silver) are typically withdrawn from circulation and hoarded, leaving in circulation only "bad" money in the form of overvalued coins (i.e., coins with very high **extrinsic** value). This principle, known as Gresham's law (from Thomas

Gresham, the 16th-century financier of King Edward I and of Queens Mary I and Elizabeth I of England), applies especially in times of economic and/or political crisis, encountered many times in Roman history.

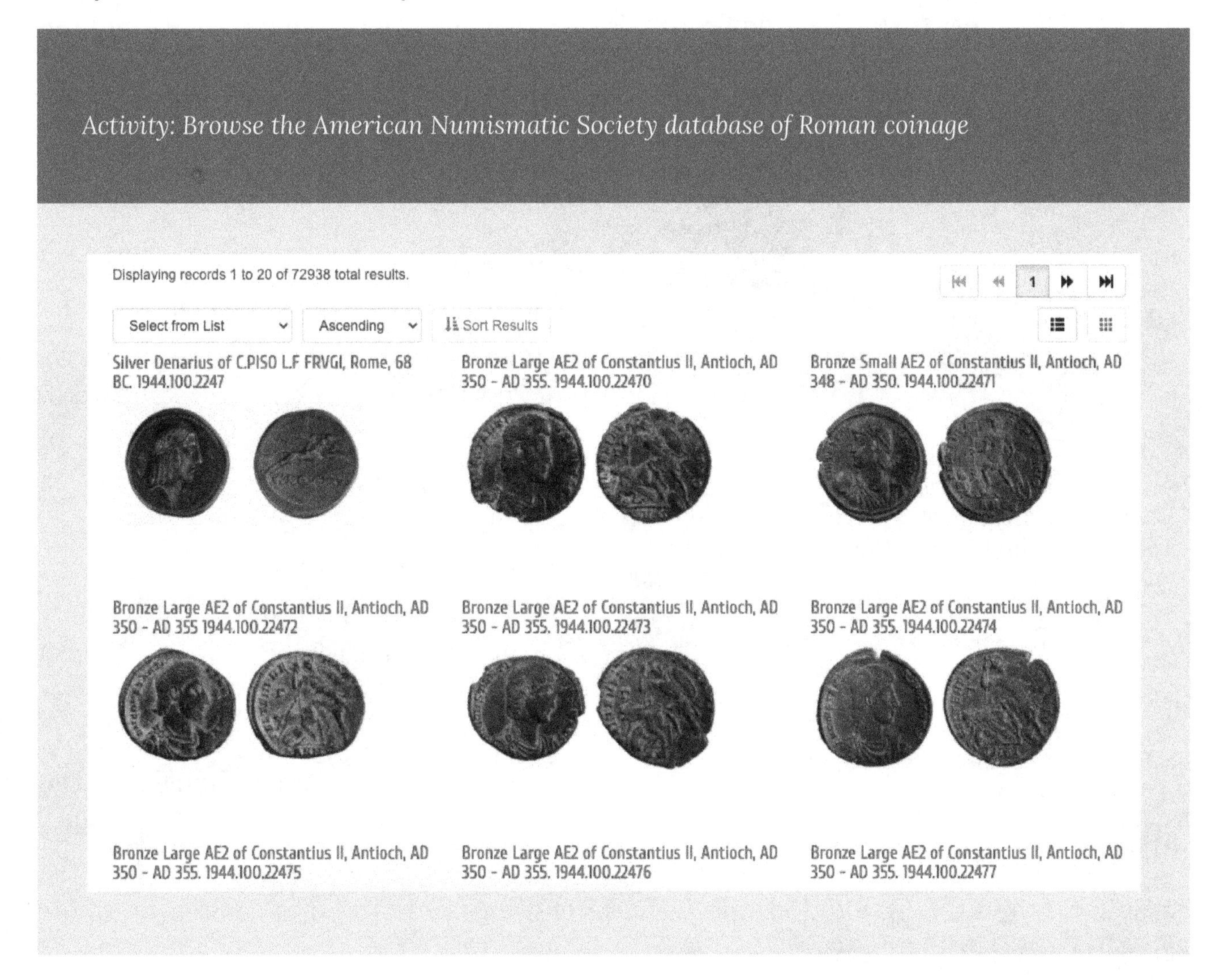

Julius Caesar first struck gold coins in very large numbers in order to pay his armies. Under Augustus, a fixed weight for gold coins and a fixed rate of exchange between gold and silver were established for the first time: the ***aureii*** were valued at 25 ***denarii*** and 41 to the pound (7.87 grams). The gold content was reduced by 4.5 percent under Emperor Nero in 64 CE; however, the rate of exchange between silver and gold did not last very long. When the silver coinage completely collapsed in the 3rd century CE, both silver and bronze were worthless. Gold and goods in kind were now the only means to keep the Roman economy afloat. In other words, in a trimetallic system, gold tends to be placed higher than the other two metals. This is in fact what brought the 3rd-century economic crisis to an end. Emperor Diocletian (284–305) established a new standard weight for the gold coin (now renamed *solidus*). That coin remained the fundamental axis of the Late

Roman and, later Byzantine Empire for the following millennium. The monetary system in both the Late Roman and the Byzantine empires was trimetallic, but its extraordinary stability over a very long period of time (more than a millennium) was possible only by ranking gold higher than silver and bronze. With a fixed rate of exchange of gold into the other two metals, the Roman and Byzantine economies received a monetary platform of recovery and growth.

Figure 6.9 Gold coin (histamenon) of the Byzantine emperor Constantine VIII (1025–1028). The emperor is shown on the reverse, while the obverse shows an image of Christ Pantokrator (All Mighty). The histamenon (from the Greek word for "standard") was a wider and thinner version of the old solidus. The coin is 25 mm in diameter, and under Constantine VIII's successor, Michael IV (1034–1041), it was minted in a slightly concave, cup-like form, which became the standard for all gold coins of the Empire, even after Emperor Alexius I Comnenus (1081–1118) introduced the hyperperon as a new, high-quality gold issue. [Wikimedia Commons.]

Conclusion

Why were silver and gold chosen for coined money? As explained in the introduction, money fulfills four major functions: means of exchange, standard of value, storage of wealth, and means of payment. However, the choice of a particular material for money depends upon certain basic criteria: durability (that material cannot ruin easily), divisibility (that material must be easy to divide into smaller denominations), portability (that material must be easy to move around), homogeneity (each unit should be the same as all others), and acceptability (people need to recognize that material easily for daily use). In addition, the material chosen needs to be of limited or stable supply, for money is in fact deferred consumption.

From the point of view of a modern engineer, however, any element in the periodic table may served as coined money, as long as it is not a gas, and is neither corrosive and reactive, nor radioactive. To be sure, applying those criteria leave about 30 eligible elements. However, since in order to be coined, a metal must also be rare, the number of elements may be reduced to only eight—the so-called precious metals: rhodium, palladium, osmium, iridium, ruthenium, platinum, gold, and silver. The first five elements were not known before the 19th century, and platinum has a melting point of about 3,000 degrees Fahrenheit, a temperature that was impossible to obtain before the first modern furnaces. That left only gold and silver for the job. Gold is the most malleable of all metals and does not readily react with air. A single gram of gold can be beaten into a sheet of one square meter, or an ounce into 100 square feet. Silver tarnishes when exposed to sulfur compounds in air or water, but has the highest electrical and thermal conductivity of all metals. The conductive properties of gold and silver may not have been either apparent, or indeed necessary to ancient moneyers, but malleability and tarnish resistance were most definitely key properties in selecting gold and silver for coined money. In various languages the very word for "money" derives from words for those two metals: *argent* in French (derived from Latin *argentum*, which means silver), *arian* in Welsh (a word that means both "money" and "silver"), *Geld* in German (a cognate of *Gold*), and *dengi* in Russian (derived from the Turkic word *tenge*, which refers to a silver coin).

Why is it important to learn about the use of precious metals—gold and silver—for coinage? The application of silver plates to achieve better wound healing is known since the Hellenistic age, and silver nitrate was used medically throughout the Middle Ages.[8] During that same period, the production of the stained glass was based on gold.

However, in both cases, such applications had little, if any, impact on the general perception of silver and gold as precious because of being coinage metals. In other words, until relatively recently, there was little, if any "technological" use of gold and silver that could compete with their primary use for the manufacture of jewelry, liturgical vessels, and coins. It is only recently that new applications have been found for gold, for example, that have dramatically shifted the emphasis away from jewels and coins. The pressure those applications put on the global resources have somewhat contributed to the rising prices of gold. Despite the fact that it is not coined anymore, and that it is used for a number of non-monetary applications, gold is still seen as a strong hedge to inflation and a store of value, since all other assets (bonds, currencies, and cash) can see their values eroded in an inflationary environment. People buy gold for investment purposes. Gold may not have any more the extrinsic value that stamping of overvalued coins gave to the metal in antiquity. However, gold is still a currency and its price fluctuates relative to other forms of exchange. This is also true for silver, which has even more industrial applications accounting for half of the annual demand of that metal. No future engineer interested in the revolutionary applications of the precious metals can therefore ignore the tensions between the intrinsic and extrinsic values of gold and silver. Still, today gold and silver possess the characteristics of a good form of money that have been established by Aristotle. However, they are not coined any more, and that opens a whole range of new possibilities for their use.

Discussion Questions

1. What is the difference, both in intrinsic and extrinsic value, between a regular coin and a bitcoin?
2. Paper money: the highest extrinsic value possible. How did people come to accept banknotes in monetary exchanges? Why were banknotes acceptable?
3. Imagine that world economies will return one day to gold and silver coins. What would the implications of such a decision be for the current (and, possibly, future) applications of gold in advanced technologies in medicine, electronics, or glass production?

Key Terms

aureus
denarius
didrachm
extrinsic
intrinsic
obverse
reverse
stater
tetradrachm

Further Reading

Harris, William V. *The Monetary Systems of the Greeks and Romans*. Oxford: Oxford University Press, 2010. http://www.worldcat.org/oclc/495598206.
Schaps, David M. *The Invention of Coinage and the Monetization of Ancient Greece*. Ann Arbor: University of Michigan Press, 2004. http://www.worldcat.org/oclc/849109350.
Wood, Diana. *Medieval Money Matters*. Oxford: Oxbow, 2004. http://www.worldcat.org/oclc/56641436.

Author Biography

Florin Curta, Ph.D. in History (1998), Western Michigan University, is Professor of Medieval History and Archaeology at the University of Florida. He has published five monographs, over 40 chapters in collections of studies, and more than 100 articles. He is also editor of six collections of studies.

Notes

1. Bernhard Laum and Heiliges Geld, *Eine historische Untersuchung über den sakralen Ursprung des Geldes* (Tübingen: Mohr, 1924), http://www.worldcat.org/oclc/868158554.
2. Jaroslav Cerny, "Prices and wages in Egypt in the Ramesside period," *Cahiers d'histoire mondiale* 4 (1954): 907.
3. Avi Gopher and Tsevikah Tsuk, *Ancient Gold. Rare Finds from the Nahal Qanah Cave* (Jerusalem: Israel Museum, 1991), http://www.worldcat.org/oclc/30622335.
4. *The World of Troy. Homer, Schliemann, and the Treasures of Priam*, Proceedings from a Seminar Sponsored by the Society for the Preservation of the Greek Heritage and Held at the Smithsonian Institution on February 21–22, 1997, eds. Deborah Dickmann Boedeker and Anna Lea (Washington: Society for the Preservation of the Greek Heritage, 1997), http://www.worldcat.org/oclc/38576319.
5. Christos Doumas, "What did the Argonauts seek in Colchis?," *Hermathena* 40, no. 151 (1991): 31–41, https://www.jstor.org/stable/23040952.
6. Robert W. Wallace, "Walwe. and .Kali.," *Journal of Hellenic Studies* 108 (1988): 203–207, https://doi.org/10.2307/632647; Gerald M. Browne, "Notes on two Lydian texts" *Kadmos* 35 (1996): 49–52, https://doi.org/10.1515/kadm.1996.35.1.49.
7. Xenophon, *Ways and Means*, trans. E. C. Marchant (Cambridge, MA: Harvard Univ. Press, 1968), 199, http://www.worldcat.org/oclc/3231818.
8. J. Wesley Alexander, "History of the medical use of silver," *Surgical Infections* 10, no. 3 (2009): 289–92, https://doi.org/10.1089/sur.2008.9941.

Steel: Carnegie and Creative Destruction

SEAN ADAMS

"Watch the costs and the profits will take care of themselves." —Andrew Carnegie

Abstract

This chapter uses the rise of Carnegie Steel as a case study to explore the social and economic context of materials innovation. In the United States during the 19th century, steel became a vital element of industrial growth, and Andrew Carnegie revolutionized its production through a system of **hard driving** at his steel mills outside of Pittsburgh, Pennsylvania. This is an example of the economic theory of **creative destruction**, in which innovation in technology and the organization of the shop floor replaces long-standing institutions and practices in the production of materials. As a result, there are both gains to society—in this case cheap steel for the construction of things like buildings and railroads, as well as drawbacks for workers and companies that tried to compete with Carnegie. In summary, innovation in the manufacture of materials can be a double-edged sword.

The President and the Weaver's Son

Franklin Pierce should have been celebrating his inauguration as President of the United States in early 1853; instead he was mourning. The New Hampshire Democrat had just won 86 percent of the electoral vote in the presidential election, even though he was a long shot to even win the nomination. Franklin, his wife Jane, and their beloved son Benny boarded a train on January 6, 1853, in Andover, Massachusetts. Less than one mile into the journey, a coupling snapped and the passenger car rolled down an embankment. Jane and Franklin suffered only slight injuries, but Benny was not so lucky. The Pierces watched horrified as the wreckage from the crash decapitated their only living son. Both parents would never be the same again; Jane brooded over the accident and even wrote a letter of apology to her deceased Benny, while Franklin never displayed the political vigor needed to unite a county rapidly coming undone over the issue of slavery.

Figure 7.1 President Franklin Pierce [Portrait by George P.A. Healy (1858), White House Collection. Wikimedia Commons.]

Although they felt their tragedy intensely, as every grieving parent would, the Pierces' experience was not a rare occurrence. In only two years, 1850 to 1852, over 900 Americans died in railroad accidents in New York State alone. Deficient parts, such as the coupling that failed the Pierce family in January of 1853, and worn-out iron rails signaled a technological challenge for American railroads, as well as a safety concern that could impact ticket sales.[1]

Figure 7.2 Young Andy Carnegie and brother Thomas, ca. 1851 [Photo from Project Gutenberg edition of Autobiography of Andrew Carnegie (2006). Wikimedia Commons.]

At the same time that the Pierces mourned Benny, young Andrew Carnegie had his own problems. His father, Will Carnegie, had been a skilled weaver in Scotland. But when a steam-powered loom opened in his hometown of Dumferline, the elder Carnegie struggled to find steady work. "Andra," he confided to his wife in 1847, "I can get nae mair work." The next year the Carnegies were on their way to America, where Will still could not find a good job. In 1855, while a gloomy President Franklin Pierce presided over Washington, Will Carnegie died, leaving Andrew to tend to his mother, his younger brother, and fend for himself.[2]

On the surface, the lives of Franklin Pierce and Andrew Carnegie appear very different. What could unite the fourteenth President of the United States and the immigrant son of an underemployed weaver? As it turns out, we can use steel to forge a link between these two stories; in fact, steel tells a great deal of the story of Industrial America. Think of it this way: had Andrew Carnegie's steel been around in 1853, the history of the Pierce family might be quite different. But even as cheap and durable steel made railroad travel safer during Carnegie's time, the societal cost—particularly to workers and to Carnegie's competitors—took an altogether different toll on the US. How this connection between steel and 19th-century Americans took hold is the focus of this chapter. By understanding the context in which steel emerged as an essential material for industrialization, we can appreciate the ways in which new materials enrich our own generation. Steel presents a case study that provides a kind of blueprint for anticipating how the integration of new materials into society can generate both positive and negative results. This case clearly illustrates the impact of materials on society.

Iron and Steel

Iron is the second-most common element on Earth, and humans have worked it into tools, weapons, and other goods for millennia. The process to convert iron ore into metal of various levels of toughness and flexibility is simple and straightforward: cook the impurities (mostly oxygen) away from the ore, leaving the base metal behind. Steel requires a bit more work to make, as it needs a particular amount of carbon—usually between one and two percent by weight—and this kind of precision requires more care than basic iron smelting.

The relationship between humans and iron remained relatively stable over thousands of years until the 19th century, when innovations in production techniques and changes in consumption patterns accelerated the knowledge and practice of iron metallurgy quite suddenly. In fact, these changes occurred so rapidly, we refer to the period as an "industrial revolution," and iron and steel played an absolutely central role in that phenomenon, particularly in the US. America became connected to steel in so many ways, that by the advent of the 20th century, they didn't think twice about its presence in their lives.

Why Steel?

In order to understand why steel is so important, we first have to dive into the history of iron. After all, steel is simply an alloy of iron and carbon, but since the common definition of steel is that it contains up to two percent carbon by weight, its history is closely related to iron. The main property that makes iron such a desirable product is its toughness—an iron plow tears through the soil much more effectively than a wood one—but at very high purity levels, iron can be too weak and ductile. Adding carbon to iron adds strength, but maintains the product flexibility; adding too much carbon makes it brittle. Steel with a well-controlled carbon content is the most desirable form of an iron carbon alloy. Steel can maintain a sharp edge quite well, which is why it has been a popular material for making swords for centuries. However, making steel was a difficult, labor-intensive process for most of human history.

Iron, on the other hand, was relatively easy to manufacture. Iron smelting is the manufacturing process that mixes iron ore, a carbon-based fuel like charcoal or coal, and a stabilizing agent such as limestone. Small furnaces were found in ancient China, Mesopotamia, and Rome, which proved that iron smelting has been used for thousands of years. Pure iron reaches a liquid state at temperatures of 1,500 degrees Celsius, which is beyond the capabilities of ancient furnaces. These early furnaces produced a product called a **bloom**, which is a mixture of pure iron and non metallic oxides. The bloom then had to be heated and hammered in order to remove the oxides. This labor-intensive process meant that iron was a useful material for weapons, small tools, and decorative items, but large-scale manufacturing was not possible.

Figure 7.3 Cutaway of a 19th century blast furnace

By the 1700s, Europeans were constructing large, tower-like **blast furnaces** that could run for months on end. These large-scale furnaces could reach temperatures sufficient to create cast iron (iron with >2.3 percent carbon), also known as **pig iron**, which has a lower melting point than pure iron. The higher temperature of the blast furnaces was achieved by forcing air up through the mixture of iron ore and coke. Coke is coal that was previously heated to reduce the sulfur content. The iron subsequently liquefied at the bottom of the furnace while the impurities in the ore bonded with the limestone or other reactants to form **slag**, a glassy-like mixture. Once the iron-maker decided the batch was ready, workers tapped the furnace (allowed the molten cast iron to drain off the bottom) and skimmed the impurities or slag off of the top. The molten iron flowed into troughs dug into the sand surrounding the furnace; workers eventually called the cooled iron "pigs" because they looked as if they were baby pigs suckling a large sow. The "pig iron" could then be reworked into iron tools, sash weights, cannonballs, plows, stoves, and other products by pouring, or "casting," the melted iron pigs into molds, thus the term cast iron. As stated, this form of iron had lots of carbon (>2.3 percent). Once the iron was smelted, blacksmiths could also use finery forges to remove most of the carbon using gas oxidation of the melt, thereby creating wrought iron. Wrought iron could then be worked into steel by adding small amounts of carbon back into the iron, a process called carburization. So, **blast**

furnaces made larger cast iron products more affordable, but cheap steel continued to elude furnace masters.[3]

Iron in the American Context

Figure 7.4 *Digging out the moulds for iron "pigs" at a blast furnace [Photo by Keystone View Co. (ca. 1905), Library of Congress.)*

Early American iron makers built their furnaces in remote locations known commonly as "iron plantations," because they required close proximity to essential resources such as a moving stream for water power, iron ore, and wood for charcoal fuel. At full capacity a good American blast furnace produced 25 to 30 tons of "pig iron" a week. This raw product made its way to specialized facilities like rolling mills, nailworks, and wireworks that produced more refined types of iron products.

By the early 1700s, British iron makers were forced to use coal in place of charcoal, because of reduced stocks of wood. This technology crossed the Atlantic to appear in American furnaces by the 1840s. The substitution of coal for charcoal reduced the cost of making a ton of pig iron by half. The use of coal also meant that iron furnaces and foundries (those smaller facilities that worked pig iron into useful tools and products) could be built closer to urban centers. As the nation grew, iron became an essential ingredient in American life. Farmers plowed their fields with an iron plow in the morning, cooked their meals in cast iron skillets at midday, stirred their fires with iron tongs to warm the chilly evening air, and then closed their windows at night using iron counterweights.

Figure 7.5 *Plow advertisement from 1879 [Wikimedia Commons.]*

Iron, Steel and Railroads

At the same time that coal became a common fuel in American furnaces and iron a common product in its households, the nation's expanding railroad network increased demand for steel. In the three decades before the Civil War, the American railroad network of the United States grew by a factor of ten; these railroads needed iron products such as rails, cars, and other railroad components. American iron producers struggled to meet this demand, and during this period of great expansion, many railroads imported their rails from Great Britain.

Some railroads tried to save on iron by using "strap" rails, which were thin iron strips that sat on top of wooden rails. Iron-starved companies could build a line quickly and cheaply in this fashion, but the strips of iron sometimes came loose and would fly upwards when a passing locomotive ran over them. These "snakehead" rails literally impaled some railroad passengers, and were part of the gruesome reality of early railroad travel. American railroads in the antebellum era (before 1860) also purchased iron products that were of uneven quality. Couplings, axles, and wheels were notorious for failing quickly—and sometimes disastrously—as the Pierce family found out so tragically. Castings made from pig iron were cheap, but the quality was uneven. Without the means of testing it for strength and flexibility, manufacturers made inferior products that often broke down at the most inopportune times.

The Work of Making Steel

Could steel be the solution? Since steel is stronger than cast or wrought iron it became the ideal material for making durable rails and other parts needed to run a railroad safely and efficiently. But at the time when the nation's rail network needed cheap steel, most of the steel made in the United States at that time was **blister steel**, which required the repeated application of powdered carbon to superheated wrought iron to add enough carbon to create steel. A skilled blacksmith then repeatedly hammered the carbon into the blade and dunked it into water to quench or cool it rapidly. This ancient art of making high-quality steel was fine for razors and swords; but it was not feasible for the mass production of large industrial products like rails.

Figure 7.6 A gruesome depiction of a railroad accident in 1856. Although sometimes caused by human error, accidents often resulted from faulty materials. [Lithograph by John L. Magee, Library Company of Philadelphia.]

Figure 7.7 An iron puddler tapping steel from the furnace [National Photo Company Collection (1919), Library of Congress.]

Another method of making steel, which also drew upon highly skilled labor, was called puddling. This method of refining involved using the heat of a coal fire to melt pig iron in a furnace while an experienced **puddler** stirred the molten iron, which burned off impurities while the purer iron formed a pool, or puddled, in the furnace. Eventually the puddler added carbon to the iron and quite literally stirred up a batch of steel. As with blister steel, there was no inherent problem with this method of making steel, other than the amount of time and skilled labor it took. The necessity of skilled labor for this intensive process imbued the workers with a great deal of control. This necessarily raised the cost of production and limited the production output.

In 1856, however, Henry Bessemer invented a process of converting pig iron into steel, which used the injection of a blast of air through molten iron. The **Bessemer Process** used an egg-shaped furnace that tilted once to accept the molten iron and then returned to an upright position so that impurities could blow out the top of furnace.

Figure 7.8 Making Bessemer Steel in Pittsburgh [Illustration in Harper's Weekly (April 10, 1886), Library of Congress.]

Who Actually Made the Steel?

In the years following the Civil War, most of the power over the steelmaking process resided with the workers. This was reflected in both wages and working conditions. For example, in 1865 the Sons of Vulcan, an organization of iron puddlers, secured a "sliding scale" for their members in which wage rates were tied to the price of iron. However, as new techniques eroded the traditional power and prestige of puddlers and other skilled ironworkers, many employees sought to organize themselves into unions. In

1876, the Sons of Vulcan combined with a number of other trade unions to form the Amalgamated Association of Iron and Steel Workers in Pittsburgh. By 1891, the Amalgamated reached its peak membership of over 24,000 workers organized into 290 local unions. They exerted a great deal of authority and helped organize what had been a fairly diffuse trade.[4]

Figure 7.9 Steelworkers at Carnegie Steel's Homestead Works, 1890 [Historical Society of Western Pennsylvania.]

Who Bought Steel?

These innovations in production were made possible by an insatiable demand for steel in American economy of the late 19th century. After the Civil War, railroads grew at a spectacular rate, until by 1890 they linked the entire continental United States with 167,000 miles of rail; 150,000 miles had been laid since 1865. In 1877 American steelmakers made 432,169 tons of *steel* rails, considerably more than the 332,540 tons of *iron* rails produced in the same year.[5] Every American lived and worked, on average, within ten miles of a rail line. This kind of a massive network required a constant influx of cheap steel, as new rail lines went in, old ones were replaced, and rolling stock like locomotives, freight cars, and passenger liners all employed steel parts for their durability and strength. This was no longer Franklin Pierce's kind of railroad.[6]

As American cities spread out across the country, they also grew up with vertical construction. The massive influx of immigrants to cities across the US made urban areas in the Northeast and Midwest swell in size. During the period from 1860 to 1880, the population of New York City grew from 813,000 to 1.2 million, and Chicago's population grew from about 112,000 to over 500,000. This kind of rapid growth meant that urban infrastructure was stretched to its limits. In Chicago, a solution came with steel construction. Between 1885 and 1895, Chicago architects developed a close relationship with Pittsburgh steelmakers to develop a "steel-skeleton" design that allowed for individual buildings to double, even triple in height over the traditional brick and mason structures. When William Jenney's Home Insurance Building was completed in 1884 with structural steel, it was the tallest building in the world at 138 feet.

Fifteen years later, New York's Park Row Building's thirty floors stretched 391 feet and helped coin a new phrase: skyscraper. As American cities grew vertically, so did the demand for structural steel.[7]

SCIENTIFIC AMERICAN

A WEEKLY JOURNAL OF PRACTICAL INFORMATION, ART, SCIENCE, MECHANICS, CHEMISTRY, AND MANUFACTURES.

NEW YORK, DECEMBER 24, 1898.

Figure 7.10 *The Park Row Building on the cover of Scientific American in 1898. [Wikimedia Commons.]*

Railroad construction showed no signs of decreasing, and the application of steel beams to the construction of buildings and bridges meant that iron and steel rails and beams commanded a large share of domestic consumption in the United States. As the industrial economy expanded, the demand for steel in the US seemed insatiable.

The Politics of Steel

Politicians knew that steel was critical to their nation's success; so American steel enjoyed a protective tariff for most of its rise, which meant that any consumer of imported iron had to pay a sizable duty. To insure this was the case, the American Iron and Steel Association, the industry's trade organization, teamed up with protectionists in the US Congress like Pennsylvania's Rep. William "Pig Iron" Kelley. Their efforts ensured that while tariff levels fluctuated in the years prior to 1900, they remained relatively high. Operating with strong tariff protection and utilizing new technologies such as the Bessemer process, the American iron and steel industry blossomed in the years after the Civil War.

Carnegie and Steel

Here is where Andrew Carnegie reenters the picture, because if you're going to talk about the rise of cheap steel in the American economy, you need to know the story of Andrew Carnegie. He was not the first person to make steel, nor would he be the last. But his method of doing so would change the course of the industrial economy forever.

By the time Andrew Carnegie immigrated with his family to the United States in 1848, steel was still costly to make, and so his newly adopted country imported a great deal of it from Great Britain. While working as a telegraph operator with the Pennsylvania Railroad, young Andy saw firsthand the American railroad industry's huge appetite for high-quality steel. After earning a small fortune in the stock market, Andy decided that he wanted to make something rather than just buy or sell things. That something was steel.

Hard Driving

Figure 7.11 The Edgar Thompson Works outside of Pittsburgh, Pennsylvania in 1891. [Library Company of Philadelphia.]

When he finished construction on his Edgar Thomson Steel Works outside of Pittsburgh in 1873, Andrew Carnegie was employing two major business strategies. The first was a tactic called hard driving, in which Carnegie worked his men and his machines to the limit. He used the newly improved Bessemer process for making steel, which required a massive furnace with a specialized lining, which Carnegie's employees wore out constantly. But he didn't care, so long as he continued to reduce the cost of steelmaking.

Carnegie Steel hired Alexander Holley to implement Bessemer steel technology, the cutting edge steel making technology at the time, at his mills. When a new and more efficient process known as **open hearth** was developed, Carnegie simply scrapped the existing equipment and made the transition to the new system. Short-term costs were no object for improved efficiency. For example, once a manager told Carnegie he knew of a rolling-mill design that could roll steel rails more efficiently. Andy ordered the existing rail mill—which was only three months old—ripped out and the new one installed.

Carnegie's managers tinkered with the process constantly—finding more and more ways to cut the cost of labor and materials. If that meant adopting new techniques, Carnegie's managers did it quickly; if it meant breaking the power of skilled workers on the shop floor, Carnegie's managers did it brutally. In the end, hard driving was very successful: the first ton of Carnegie steel cost about $56 a ton to make; by 1900 the cost was down to around $11.50, a dramatic 79 percent cost reduction in about 27 years.

Hard driving is a great example of the more general concept of improving **throughput**: essentially a measure of the speed and volume that the flow of materials has through a single plant or works. A high rate of throughput—which managers usually measured in terms of units processed per day—became the critical criterion of mass production. It is an important measure of productivity that many businesses sought, to emulate the success of steel.

Key Concept: Bessemer Steel

In 1856, Henry Bessemer introduced, to London's Royal Academy of Science, a revolutionary process of converting iron into steel, which used the injection of a cold blast of air through molten iron in an egg-shaped furnace—called a "converter"—that transformed iron into steel in a matter of minutes.

Although accounts vary as to who actually "invented" the Bessemer process, we do know that the first working converter in the United States appeared in Troy, New York, in 1864. Soon other firms like the Cambria Iron Works and the Bethlehem Iron Company quickly adapted Bessemer converters to their plants, and Carnegie Steel eventually turned the process into a finely tuned machine.

Carnegie was obsessed with reducing costs via innovation and efficiency, and with his strict adherence to hard driving, his steel factory replaced its Bessemer Converters for the open-hearth process, which is similar to the way steel is made today.

There are only a few Bessemer converters remaining in the US, and most of them are museum pieces.

The Carnegie Legend

By implementing new technology and cost accounting, Carnegie saw one blast furnace increase annual output from 13,000 tons to 100,000 tons. British steelmakers, who had previously dominated the world market in steel rails, couldn't understand this process and considered it reckless. For example, Carnegie would hard drive a furnace until the lining was completely shot, then he would simply replace his furnaces about once every three years. British furnaces, in contrast, lasted 12 years on average. One British visitor bragged to Carnegie that back home they had equipment they had been using for 20 years. Carnegie reportedly responded, "And that is what is the matter with the British steel trade." Carnegie streamlined the production process within the plant. Once the blast furnace formed the molten steel into ingots, they were rushed to the rolling mill and made into rails. Another British observer said that he would like to sit on an ingot for a week and watch that mill operate. A manager told him that if he wanted an ingot cool enough to sit on, he'd have to send to Britain for it. When steelmaking changed from the Bessemer system to **open-hearth** production (a method of steelmaking that relies more on controlling the chemistry), Carnegie removed the Bessemer converters and put in open-hearth furnaces. The end result was that Carnegie consistently slashed prices and undersold competitors.[8]

Vertical Integration

Figure 7.12 Making coal into coke in "beehive" ovens. [Pennsylvania State Archives.]

Figure 7.13 Andrew Carnegie, the Hard Driver, ca. 1878. [Photo from Project Gutenberg edition of Autobiography of Andrew Carnegie (2006). Wikimedia Commons.]

The second major strategy that Carnegie employed was something business historians called **vertical integration**. Rather than buy the ore, Carnegie bought mining land in Minnesota's rich Mesabi Iron Range, along with a small fleet of vessels to transport it to Pittsburgh. Rather than buy the best fuel for steelmaking—a refined version of coal called coke (coal with reduced sulfur content)—Carnegie acquired vast coalfields, as well as the beehive ovens that made coke, when he brought Henry Clay Frick, the ruthless coal baron, into Carnegie Steel. By 1900, his company controlled every aspect of steelmaking, from the time the ore left the ground to the time it appeared in the form of a steel rail.

Andrew Carnegie's many innovations make him a great example of an entrepreneur in America's Industrial Age. He revolutionized the steel making process, in order to undercut his competitors, increase his share of the steel rail and construction beam markets, and drive any potential competition out of business. As a nation, America benefited from Carnegie's cheap steel, but there were costs as well.

Creative Destruction

As steel transformed the American landscape in both the city and the country, Andrew Carnegie reshaped both its method and marketing. What would be the impact of these changes? Why are they important?

In 1942, the economist Joseph Schumpeter coined the phrase **creative destruction** to describe the process of industrial change, particularly in the face of entrepreneurial activity and the incorporation of new technology. Schumpeter was trying to explain how the market economy drives change that benefits society, but at the same time can destroy established ways of doing things. In his book entitled *Capitalism, Socialism, and Democracy*, he described a process of "industrial mutation" in which a new way of producing goods is "incessantly destroying the old one, incessantly creating a new one. This process of Creative Destruction is the essential fact about capitalism." [9]

Frick and Homestead

Henry Clay Frick (1849–1919) had built a business empire in the coalfields of Southwestern Pennsylvania. Frick produced the best "coke," a fuel made by baking coal in beehive-shaped ovens. After consolidating his power in the coking fields of Pennsylvania, Frick joined Carnegie Steel as a partner in 1889.

Frick's iron-fisted tactics in breaking labor unions in his coking operations influenced his approach to the workers' strike at Carnegie's Homestead Works in 1892. He locked them out and prepared for a lengthy struggle.

Frick was a loathed figure in the American public eye, but this changed in July of 1892 when an anarchist, Alexander Berkman, tried to assassinate Frick. Berkman burst into his office, shot him twice in the neck and stabbed him four times. Frick survived this attack and returned to work after only a few days. Berkman and his anarchist ties discredited the strike among middle-class Americans, even though the Amalgamated had nothing to do with the assassination attempt.

Carnegie and Competition

Any firms who tried to cut into Carnegie's market found themselves in a tough spot; Carnegie would drive the price of steel low enough to put the upstarts out of business, only to raise the price back up once the coast was clear. Some immediate benefits of creative destruction were the increased supply of cheap steel rails and structural beams, but in rearranging the way America did business, Andrew Carnegie's process left wreckage in his wake.

Carnegie Steel's price leadership limited smaller firms to niche markets such as structural steel, wire, wire nails, rods, and hardware. So it is unfair to say that Andrew Carnegie enjoyed a monopoly on steel production or even an oligopoly. Instead, he drove out competitors in large industrial markets and focused on achieving **economies of scale**—that is, the reorganization and expansion of the production process so as to reduce costs. In order to succeed with this strategy, Carnegie needed absolute control over the process of making steel, from the raw materials to the finished product.

The Homestead Strike

Because of his working-class background, Andrew Carnegie liked to portray himself as a friend of his workers. In 1889, his workers went on strike at his steel plant in Homestead (outside of Pittsburgh) and Carnegie settled by giving them a contract that set their wages higher than those at neighboring steel mills and negotiated with the union, the Amalgamated Association of Iron and Steel Workers. Homestead was huge—12 mills employing 3,800 men with a town of 11,000 surrounding it.[10]

The integration of new technologies like the Bessemer converter or the open-hearth furnace threatened the Amalgamated's power. In prior labor conflicts, steelworkers had been able to gain some concessions from their employers like the sliding scale. But when faced with the system of hard driving, the Amalgamated struggled to retain control over the shop floor. This all came to a head in Homestead, Pennsylvania, during a famous strike.

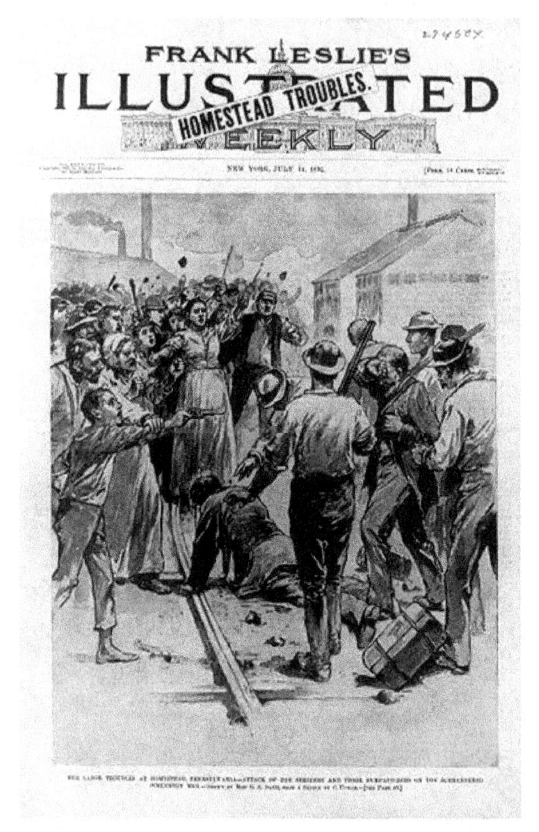
FRANK LESLIE'S
ILLUSTRATED
WEEKLY
HOMESTEAD TROUBLES.

Figure 7.14 The Homestead Strike was front-page news in America's magazines and newspapers in 1892. [Illustration by G.A. Davis, from a sketch by C. Upham, in Frank Leslie's Illustrated Weekly (July 14, 1892). Library of Congress.]

In 1892, when Homestead workers walked off the job, Carnegie left the country to travel to Scotland and left the dirty work of union busting to Henry Clay Frick. Frick announced that Carnegie Steel would only deal with men on a one-to-one basis, and the workers walked out. Frick then set up "Fort Frick," a 12-foot-high fence topped with barbed wire, and hired 300 Pinkerton Agency detectives as guards. On July 2, he shut Homestead down to lock out the union. When the Pinkerton guards arrived on a barge on the Monongahela River on July 5, the workers met the Pinkertons with gunfire, rock projectiles, and a small cannon. After 12 hours, nine strikers and seven detectives were killed. The National Guard was called out, the Pinkertons were allowed to leave, and Homestead opened with non-union workers. The strike went on until November, when the Amalgamated finally gave up and called it off; the union leaders were fired from Homestead and Frick sent a wire to Carnegie in Scotland: "Our victory is now complete and most gratifying. Do not think we will ever have any serious labor trouble again." In fact, the steel industry operated without organized labor for the next four decades.[11]

In many ways, the Homestead Strike represented the culmination of years of creative destruction at Carnegie Steel. As a result of hard driving and union busting, Andrew Carnegie had full control over his shop floor. As a result, workers found no choice but to take the wages Carnegie Steel offered and follow the managerial directives that Carnegie Steel wanted. Steelworking would never be the same.

U.S. Steel

In 1900, the financier J.P. Morgan was tired of competing with Carnegie Steel, which kept beating Morgan's prices in steel-making ventures. So he bought out Carnegie Steel and formed U.S. Steel, the world's first corporation capitalized at over $1 billion—precisely $1.4 billion. Andrew Carnegie's personal take in the U.S. Steel deal was over $200 million—billions in today's dollars—and he was perhaps the world's wealthiest man. He could not have done it without the creative destruction of the steel industry. Andrew

Carnegie's approach to industrial manufacturing, from hard driving and vertical integration to creative destruction, are all factors that make his story fascinating and important to the rise of steel.

The Legacy

What does this case study in steel tell us about materials? It tells us that innovation can be a double-edged sword, having both positive and negative impacts on societies and the world. On one hand, Carnegie's cheap and durable steel made railroads more affordable and safer. The story of Franklin Pierce's tragic loss might have been quite different had Carnegie Steel been available fifty years earlier. On the other hand, Andrew Carnegie's rearrangement of the shop floor, his ruthless competitive business practices, and his willingness to break the power of labor organizations like the Amalgamated, shattered lives. His own father, Will Carnegie, experienced the destructive side of creative destruction right before moving his family to America. Steel changed the United States by allowing the nation to literally expand both outwards (railroads across the nation) and upwards (skyscrapers) while some Americans, like those workers at Homestead, paid a high price for cheap steel.

Discussion Questions

1. Do you think that Andrew Carnegie's firsthand experience with the mechanization of his father's workplace might have affected his strategy in steelmaking?
2. Do the benefits to society of having an Andrew Carnegie reshaping the steel industry outweigh the costs to workers and competitors?
3. Can you think of other examples in which "creative destruction" transformed the production of a material?
4. What materials might revolutionize contemporary life if they were suddenly made abundant and cheap?

Key Terms

hard driving
creative destruction
bloom
slag
blast furnace
pig iron
vertical integration
economies of scale
throughput
open-hearth
puddler
Bessemer Process
blister steel

Author Biography

Sean Adams is the Hyatt and Cici Brown Professor of History at the University of Florida in Gainesville, where he teaches courses in the history of American capitalism, the global history of energy and 19th-century U.S. history. His most recent book on energy transitions and home heating is entitled *Home Fires: How Americans Kept Warm in the 19th Century* (Johns Hopkins, 2014), and it explores the roots of America's fossil fuel dependency during the Industrial Revolution. He is also the author of *Old Dominion, Industrial Commonwealth: Coal, Politics, and Economy in Antebellum America* and a three-volume anthology entitled *The American Coal Industry*, 1789–1902, along with numerous articles, reviews, and book chapters.

Further Reading

Knowles, Anne Kelly. *Mastering Iron: The Struggle to Modernize and American Industry*, 1800–1868. Chicago: University of Chicago Press, 2013. http://www.worldcat.org/oclc/839276664.

Krass, Peter. *Carnegie*. Hoboken, NJ: Wiley and Sons, 2002. http://www.worldcat.org/oclc/50143490.

Standiford, Les. *Meet You in Hell: Andrew Carnegie, Henry Clay Frick, and the Bitter Partnership That Transformed America*. New York: Three Rivers Press, 2005. http://www.worldcat.org/oclc/1150266762.

Notes

1. Michael Holt, *Franklin Pierce* (New York: Times Books, 2010), 50, http://www.worldcat.org/oclc/435711537; Mark Aldrich, *Death Rode the Rails: American Railroad Accidents and Safety, 1828–1965* (Baltimore: Johns Hopkins Univ. Press, 2006), 19–20, http://www.worldcat.org/oclc/213305539.
2. Harold Livesay, *Andrew Carnegie and the Rise of Big Business* (Boston: Little, Brown and Company, 1975), 7–18, http://www.worldcat.org/oclc/607827467.
3. Peter Temin, *Iron and Steel in Nineteenth-Century America: An Economic Inquiry* (Cambridge, MA: MIT Press, 1964), 57–62, https://archive.org/details/ironsteelinninet00temi.; Robert Gordon, *American Iron, 1607–1900* (Baltimore: Johns Hopkins, 2001), 90–124, https://muse.jhu.edu/book/72153.
4. David Brody, *Steelworkers in America: The Non-Union Era* (Cambridge, MA: Harvard Univ. Press, 1960), 50–57, http://www.worldcat.org/oclc/6707478.
5. S.H. Finch, "The Iron Industry in its Relation to Railways," *American Railroad Journal* 58 (February 1885): 353–54, https://archive.org/details/5088829_58.
6. Walter Licht, *Industrializing America: The Nineteenth Century* (Baltimore: Johns Hopkins Univ. Press, 1995), 82–83, http://www.worldcat.org/oclc/655201338
7. Thomas Misa, *A Nation of Steel: The Making of Modern America, 1865–1925* (Baltimore: Johns Hopkins Univ. Press, 1999), 45–89, http://www.worldcat.org/oclc/540692649.
8. Harold Livesay, *Andrew Carnegie*, 109–23.
9. Joseph Schumpeter, *Capitalism, Socialism, and Democracy* (New York: Harper and Brothers, 1942), 83, http://www.worldcat.org/oclc/30488029.
10. Paul Krause, *The Battle for Homestead: 1880–1892* (Pittsburgh: Univ. of Pittsburgh Press, 1992), 245–51, http://www.worldcat.org/oclc/608933009.
11. David Nasaw, *Andrew Carnegie* (New York: Penguin, 2006), 428–38, http://www.worldcat.org/oclc/255688279.

Aluminum: Alcoa and Anti-Trust

SEAN ADAMS

"There is no such thing as a good monopoly." —Learned Hand

Abstract

This chapter uses Alcoa to tell the story of aluminum and antitrust. Although aluminum is quite common now, it was a very difficult metal to refine before 1888. One American company, Alcoa, was responsible for the rise in the application of aluminum to various markets. In order to increase production and profits, Alcoa grew in both size and scope, as did many American businesses during the early 20th century. Eventually, this company controlled two-thirds of the world's supply. But as Alcoa emerged as a big business, federal policymakers considered it a dangerous threat to competitiveness in aluminum production. Using the doctrine of antitrust, the United States successfully knocked Alcoa down from its lofty place in worldwide markets in 1945. In doing so, those policy makers believed they had struck a blow for competitiveness, but the larger question arises: is a company that dominates the production of a single material bad in and of itself?

Introduction: Capping the Monument

Aluminum is the most common metal in the earth's crust. It has amazing properties—it is resistant to corrosion, its low density means that it is one of the most lightweight metals, and it has a relatively low melting point, which means that it is easy to cast. We encounter aluminum nearly every day: it is in our beverage cans, our automobiles, planes, and in many of our kitchens. In fact, you might say that aluminum is the most commonplace metal of all. But this was not always the case.

Take, for example, the story of the Washington Monument. America had been planning to build some sort of memorial structure to commemorate the life of their first president, George Washington, since the 1780s. But, as with many monuments, the construction was subject to great controversy. One of the designs, for example, fell by the wayside when it came out that the Pope had donated stone from an ancient Roman temple for its construction. When anti-Catholic nativists found out about this, they slowed construction on the project in the 1850s.

Figure 8.1 Capping the Washington Monument in December, 1884. [Engraving by S.H. Nealy, in Harper's Weekly (Dec. 20, 1884). Library of Congress.]

Two decades later, construction resumed with a new design: a massive obelisk with simple, clean lines that formed a point at the top. Designers thought that a metallic pyramid on top could help with lightning and keep the edges sharp and clean. After considering bronze or copper, they decided instead to use a metal known for its attractive properties of conductivity and durability—aluminum. But they did not come to this decision because aluminum was cheap. In fact, this metal was quite expensive. In the 1880s, aluminum was about $16 per pound, or more than some American workers made in a week's work, and the eight-inch-high, cast-aluminum pyramid cost a whopping $225—over $5,500 in today's prices. The idea of using aluminum signaled the high value, not the ubiquity, of this metal. When it was completed, the Washington Monument was the tallest man-made structure in the world, standing at 555 feet, and it was topped with aluminum.[1]

Why is Alcoa Important?

At the dedication of the Washington Monument, aluminum was an exotic and expensive material. Casting the pyramid took several tries, and the manufacturer was worried that he might have use an aluminum-bronze alloy instead of pure aluminum. In fact, pure aluminum was so rare it really was more of a luxury good than anything else. Some of the royal families of Europe, for example, used aluminum dinnerware as a sign of their elite status.

The idea that an American company might unlock the secret of mass producing aluminum and find hundreds of applications for its use seems like a good one to us today; everyone likes technological innovation and efficiency. But what if this company grew so proficient at manufacturing and marketing aluminum that it controlled nearly two-thirds of the world's supply? Is it acceptable to have a single firm hold that kind of market share? Should government intervene to restore competition to that industry? Those were the questions faced by a company called the Aluminum Company of America in the years following World War I. This firm—later renamed Alcoa—was so good at making and selling aluminum they nearly cornered the market. But were they too successful for their own good? That's why the story of aluminum is wrapped up with the idea of **antitrust**, or the notion that a company that holds a huge market share operates as a **monopoly**. So why was Alcoa in such trouble?

Figure 8.2 An Alcoa aluminum press in operation. Such facilities require massive capital investments, which makes competition in aluminum production an expensive strategy. [Library of Congress.]

This chapter will use the story of Alcoa up to 1945, when the federal government reduced the company's market share through an antitrust case. This is an important example of how the political and economic context of a material can be critical in determining how it was made, what markets it serves, and how it becomes an important part of everyday life. Aluminum was not common at all before the 20th century, but Alcoa set out to change all of that. Along the way, they became almost *too* good at making and marketing aluminum. How is that possible? And why should we be concerned if a manufacturer is dominating markets, so long as that material is cheap and abundant? To understand this question we need to understand the notion of antitrust, and how Alcoa played a pivotal role in this uniquely American phenomenon.

The Origins of Big Business

Alcoa had its origins in the late 19th century, at a time when big businesses began to dominate markets. Because of technological and organizational innovation, increasing efficiency in production, and ruinous competition, prices fell steadily from the end of the

Civil War to the mid-1890s. So the only way firms could control prices in the 19th century industrial economy was to control market share. Basically, companies became bigger and bigger in order to become price-makers and not price-takers; in other words, they didn't want to leave their business to the whims of the marketplace. Instead, they sought to control their own economic strategies by growing in size and in scope, dominating the market for their goods, and setting their own prices.

These large industrial firms tried to centralize production and cut costs. There were two main goals. First, large firms built larger and larger factories in order to capture what historians call **economies of scale**. Basically, before the Civil War, most industries were subject to what economists referred to as "constant returns to scale." This means that although a bigger factory might allow the production of more goods, the costs per unit were roughly the same; for example, if you put 100 looms in a factory, they wouldn't necessarily produce cheaper cloth per unit than a single loom. After the Civil War, technological and organization changes brought "economies of scale," which means that a large, expensive plant could produce goods more cheaply on a per-unit basis than small producers. Take, for example, flour mills and oil refineries: usually massive capital outlays are required to build such plants. Second, managers focused on increasing **throughput** within those factories. Throughput is basically a measure of the speed and volume of the flow of materials through a single plant or works. A high rate of throughput—which managers usually measured in terms of units processed per day—became the critical criterion of mass production. If a company did well on these measures, it likely was a large corporation with a huge physical infrastructure—a great departure from the small-scale businesses of earlier in the 19th century.

Checking the Trusts

Economists call a system in which a few large corporations dominate the marketplace "oligarchic." Historians say that the Era of Big Business saw the rise of the "trusts," which refers to Standard Oil's legal maneuver to control over 90 percent of the oil refining business in the US economy via a system in which companies surrendered stock certificates—and control of their companies—to John D. Rockefeller via his Standard Oil Trust. Many Americans thought that the government should do something about the unchecked power of these trusts in industries like oil, sugar, beef, steel, and even whiskey. After all, if a company has a commanding market share, who is to say that they won't jack prices up to monopolistic levels? This kind of control seemed undemocratic and, to many voters, un-American. So in 1890, Congress passed the **Sherman Antitrust Act**, which made "restraints of trade" illegal and aspired to break up trusts that undermined the public interest. But the statute was vague in defining these principles.

Figure 8.3 This cartoon uses baseball as a metaphor for how the "trusts" were perceived as dominating businesses and hurting Americans. [Drawing by Frederick Opper (ca. 1901). Library of Congress.]

What is a "restraint of trade," and how do you find it? The law didn't provide any guidelines or examples, so enforcement of the law, by default, fell in the hands of the federal government to determine what a monopoly was and what wasn't.

So how do you enforce a law that is so vague? Actually, the Department of Justice first used the Sherman Act against unions. Of the first 10 cases tried under the legislation, five were against unions that supposedly acted in "constraint of trade." In 1895, the Supreme Court seemed to get the ideal opportunity to break up a trust. The American Sugar Refining Company acted in sugar in much the same way that Standard Oil did for oil. But worse, they controlled about 98 percent of market share by the 1890s. But even though the sugar trust dominated the American market, the Supreme Court, in *United States v. E. C. Knight Co.* (1895), made a very strict—and to be honest, very dubious—distinction between the Sherman Antitrust Act's applicability to commerce and its applicability to manufacturing. The Court argued that antitrust laws really only applied to commerce, not manufacturing, and unless the American Sugar Refining Company built a factory that literally straddled a state boundary, the issue was for state courts, not the Supreme Court, to decide. Since the four major sugar refineries owned by the American Sugar Refining Company were in Pennsylvania, it was a matter for the Pennsylvania courts to decide.[2]

Big Business Gets Even Bigger

The way that American courts enforced antitrust meant that firms couldn't form any kind of informal organization, like a cartel or trust. So **horizontal integration** offered the best strategy: you simply acquire your competitors and increase your market share. That way you create major barriers to your potential competitors because they'll have to catch up in the race to achieve economies of scale. To pursue this strategy, many firms used mergers or the outright hostile acquisition of other firms. The first major merger wave in American history occurred during the years 1895 through 1904. Over this stretch of nine years, more than 2,000 previously independent firms disappeared. In 1899, there were about 1,200 recorded mergers—pretty huge considering that there were fewer than 100 mergers in 1896, fewer than 400 in 1900, and then back to fewer than 100 in 1904. The firms that emerged from this merger movement often dominated markets and continued to expand in size through both horizontal and vertical integration.

Here's an example: In 1898, three regional companies—New York Biscuit, American Biscuit and Manufacturing, and the United States Baking Company—joined to form the National Biscuit Company. The directors of the new firm decided to embark upon a two-pronged strategy—centralize production through buying out competition and integrate forward to the customer. So after 1900, National Biscuit attempted to capture economies of scale through the consolidation of production facilities and increases in throughput. They also tried to develop specific brand names, like "Uneeda Biscuit," and blitzed consumers with increased advertising. With this strategy, National Biscuit kept unit costs low and created major barriers to entry for new competitors, who were limited to a few firms structured like National Biscuit, but usually operating on a regional scale. Many industries had their version of National Biscuit by the early 20th century: U.S. Steel, American Tobacco, American Bell, and the International Paper Company.

Figure 8.4 As big businesses attempted to capture market share, they often mass-produced products meant to insure consumer loyalty. [Ad by National Biscuit Company (1899). Wikimedia Commons.]

Aluminum Becomes Cheap

So what does this all have to do with Alcoa? Well, at about the same time that the Sherman Act appeared and American businesses started to grow so rapidly, there were changes in the aluminum business. Charles Martin Hall, in Oberlin, Ohio, was experimenting in the woodshed of his kitchen and developed an inexpensive way to smelt aluminum. In 1888, he filed a patent for his discovery and organized the Pittsburgh Reduction Company, which in 1907 was renamed the Aluminum Company of America, later shortened to just Alcoa. Hall was initially excited about his breakthrough, until he learned that Paul Héroult of France had discovered the same thing at virtually the same point in history. Héroult got a French patent, but never really spun his aluminum process into a commercially viable endeavor.

Figure 8.5 Charles Martin Hall in the 1880s. [Wikimedia Commons.]

The use of an electric current is the best way to reduce aluminum oxide back to aluminum. However aluminum oxide is an insulator. The **Hall-Héroult Process**, as it is called, smelts aluminum metal by passing an electric current through a solution of aluminum oxide (obtained from a mineral called Bauxite) mixed with a substance called cryolite, which is sodium aluminum fluoride. The cryolite melts at a much lower temperature than aluminum oxide, and the melt is electrically conducting. In addition, aluminum oxide will dissolve in the molten cryolite, enabling electrical reduction of the aluminum oxide to metallic aluminum. After running the current through this solution, pure aluminum begins to form on the end of a graphite rod. Initially, this process made very small amounts of aluminum, but once larger pots and a steady source of electricity could be applied, this once very difficult task of smelting aluminum became commercially viable. [3]

Selling Aluminum

When the Washington Monument was capped with an aluminum pyramid in 1884, it was about $16 per pound—that's nearly $400 per pound in today's prices—but Hall's process initially reduced that cost in half, and by 1900 his company could make aluminum for about 33 cents per pound. Aluminum very quickly evolved from an exotic metal to a lightweight material that had many potential functions. At first Alcoa sold mostly ingot and sheet aluminum, and soon moved "downstream" towards the consumer and made new goods. Teakettles and utensils, for example, became part of their lineup when they acquired utensil manufacturers. Aluminum's non-corrosive properties allowed the company to sell sheeting and tubing as well. And by 1908, they were fabricating wire and moved into the utilities market. As the automobile market developed, aluminum was a natural fit because of its lightweight alloys.

Alcoa grew very rapidly: from 1900 to 1914, the company's capital surged from $2.3 million to more than $90 million. During that time, the firm's leaders, Arthur Vining Davis and Alfred Hunt, embarked upon a policy of rapid growth in order to achieve economies of scale. They built a massive facility to generate electricity at Niagara Falls and moved beyond Pittsburgh to build large smelting facilities in New York State, Tennessee, and Canada. In order to improve throughput, Alcoa acquired bauxite mines in Arkansas and built a refining plant in East St. Louis to create alumina (aluminum oxide), the material that the Hall-Héroult Process used in order to refine aluminum metal. World War I was particularly good for business, as Alcoa increased production from 109 million to 152 million pounds; wartime applications took up to 90 percent of the firm's production. But government officials were wary, and in 1917 the War Industries Board accused the company of unfair practices when they charged a bit more for aluminum canteens than the market price. Scrutiny of Alcoa—and talk of antitrust proceedings—began to increase.[4]

Figure 8.6 Example of a World War 1 aluminum canteen

The Antitrust Problem Starts

Part of the problem was that Alcoa was coming of age during the time that the Federal Trade Commission (FTC) and the Clayton Act appeared in 1914. The FTC was created as an independent commission to enforce antitrust laws. It had five members who were elected to seven-year terms in order to isolate them from political influences. The FTC supposedly had broadly granted investigatory powers and could theoretically order businesses to stop a particular action if members thought the firm was acting in violation of antitrust laws. The Clayton Act shored up antitrust laws by making certain practices illegal. It did not allow contracts between firms that restricted firms from doing business with competitors, and it did not allow price discrimination in the effort to limit competition. For labor unions, the Clayton Act was important because it ruled that unions were not illegal combinations in the constraint of trade (remember that this is what the Sherman Antitrust Act was originally used to enforce). All this shoring up of antitrust agencies meant that Alcoa's executives had a great deal to worry about as they grew their market share during World War I.

The Southern Aluminum Company, centered in Badin, North Carolina, and backed by French investment capital, attempted to compete with Alcoa in markets across the American South. This gambit failed—Alcoa was too strong by this point—and the company's Board of Directors voted to sell all of their assets. In the era of American Big Business, this should have been a classic case of merger and acquisition, but Alcoa walked cautiously in this case. And although it bought the assets of Southern Aluminum Company in 1915, Alcoa asked for clearance from the FTC. This process was called "advance advice," and it was one potential way for U.S. policymakers to regulate the growth of industrial firms.[5]

Nevertheless, Alcoa had become an effective monopoly. Arthur Varning Davis admitted as much when he testified before the War Industries Board in 1918. "I suppose it has always been our aim to foster this industry," he told government officials. Alcoa considered itself to be the "father as well as the creator of this industry," and in regards to competition, "it has always been our conception that the stability of price was the basis on which to build the industry."[6]

Finding Markets for Aluminum

In the 1920s and 1930s, Alcoa began aggressively expanding its product line and aluminum became a commonplace metal in American industries and in the home. There were lightweight window frames, decorative railings, and all sorts of goods that took advantage of aluminum's special properties. In products that needed to be light, for example, aluminum could replace iron or copper. The virtues of aluminum's resistance to corrosion also became a selling point for many products. In 1928, an Alcoa ad bragged that "the transforming power of aluminum paint" could transform dingy industrial villages into modern towns.[7]

Figure 8.7 The decorative doors leading into the Aluminum Research Laboratory in New Kensington, PA. They are made out of, you guessed it, aluminum. [Library of Congress.]

In 1930, Alcoa built the Aluminum Research Laboratory, in which the company enlisted full-time research scientists at New Kensington, Pennsylvania, to develop new alloys and products and to improve the smelting and refining process. The scientists there worked on plastics, stainless steel, nickel alloys, and magnesium. They not only worked on aluminum products for contemporary times, but also tried to anticipate alternative technologies that might replace aluminum. Basically, Alcoa paid very smart engineers and scientists to think of ways to reduce the cost and promote the use of aluminum. It was about as close to "pure" research as one could find in the private sector, and Alcoa's commanding market share shielded the Aluminum Research Laboratory's staff from the pressure to develop new products immediately.

The Federal Government Acts

But even though it fostered a corporate culture of size and stability, Alcoa's dominance over aluminum markets created problems for the firm. In 1938, Alcoa celebrated its 50th anniversary while preparing a defense against an antitrust lawsuit. The Justice Department wanted to "create substantial competition in the industry by rearranging the plants and properties of the Aluminum Company and its subsidiaries under separate and independent corporations." During the New Deal Era of the 1930s, American policymakers became more and more concerned that large industrial corporations were dominating the economic landscape. Even though Alcoa was not rapacious in dealing with its competitors, the Department of Justice argued that its size and market share alone made it an appropriate target for antitrust proceedings. The case lasted 176 days, Alcoa spent $2 million defending itself, and, although the case was undecided when Japan bombed Pearl Harbor in late 1941, it continued during the war. One of Alcoa's executives complained, "If we are a monopoly, it is not of our choosing." Despite the lawsuit, Alcoa became heavily involved in the war effort, producing 3.5 billion pounds of aluminum that went toward the manufacture of 304,000 airplanes. Many of these airplanes used alloys that Alcoa had developed in their laboratories.[8]

Figure 8.8 During World War II, Alcoa helped build the American bomber fleet. Not all of them were named after aluminum like this one, but all of them utilized aluminum parts. [Florida State Archives.]

Nonetheless, the government still pursued the case and in 1945, on appeal, a Circuit Court justice with the greatest legal name of all time, Learned Hand, gave the verdict. He argued that there was no such thing as a "good monopoly" and that there was no way that any company could achieve a monopoly share simply through efficiency and good business practice. So, in the postwar years, Alcoa saw many of its facilities sold for pennies on the dollar to its competitors, Reynolds Aluminum and Kaiser Aluminum. In 1950, the Court set Alcoa's market share at 50.85%, which, at least in their eyes, put them in a more competitive relationship with Reynolds (30.94%) and Kaiser (18.2%).[9]

A Post-Alcoa World

This application of the Sherman Antitrust Act really had a major impact on American business. If Alcoa could be broken up because of its market share alone, then other firms might be wary of achieving economies of scale and improving throughput in order to repeat Alcoa's success. In the postwar era, many large industrial firms chose to grow in a completely different fashion. Whereas Alcoa sought to gain control over the production of a single material, many postwar firms sought to grow in completely different ways.[10]

For example, take the story of Textron, which began as a textile business. Its owner, Royal Little, was good at making textiles, but after the Second World War, the textile industry was a volatile one that suffered immensely from market swings. Little knew that he wasn't going to dominate the textile industry; even if he did, in the post-Alcoa antitrust environment, the Supreme Court would probably come after him. Beginning in 1954, Little devised a new strategy for Textron. He began acquiring small- and intermediate-sized firms at a rate of about two per month, and he was borrowing heavily to do it. In 1956 alone, Textron purchased firms that made cement, aluminum, bagging, plywood, leather, and Hawaiian cruise ships. Little also began to sell off Textron's unprofitable divisions—many of which were in textiles—and by 1963 he had sold the last of Textron's textile plants. By 1968, Little was in semi-retirement and Textron had revenues of $1.7 billion and earnings of $76 million. It was number 49 on *Fortune Magazine*'s list of the 500 largest companies—even ahead of Alcoa, which had dropped to number 56!

The other example of the move away from Alcoa's growth strategy is the story of International Telephone and Telegraph, or ITT. This firm was founded as a small telephone company in Puerto Rico and Cuba in the 1920s, and it was moderately successful when Harold Geneen took it over in 1959. After that point, ITT began to expand rapidly. It acquired Avis in 1965, Cleveland Motels in 1967, and Pennsylvania Glass and Sand, Continental Baking, and Sheraton Hotels, all in 1968. You can see where this is headed. But ITT was even more expansionist than Textron: By 1970, ITT had 331 subsidiaries and 708 subdivisions. It operated in 70 countries, had 400,000 employees, and sales of $5.5 billion—and was ranked number 21 on that same 1968 Fortune 500 list! Both ITT and Textron proved to the business world that you could get big fast—and you didn't even have to be particularly innovative in what you produced. With some creative accounting, gutsy acquisitions, and stock swaps, Textron and ITT became giants over the course of a decade. And they did it with a completely different strategy than Alcoa. Rather than make one thing exceptionally well, conglomerates sought to make profits across many markets and avoid antitrust problems altogether.

The Beer Can Barons

Don't weep too much for Alcoa in the postwar era. From 1946 to 1958, its gross revenues tripled up to $869 million. One writer in 1955 called it Alcoa's "splendid retreat" from monopoly. By 1952, moreover, aluminum had passed copper in civilian consumption and now is second only to iron. In 1977, the NASA's Space Shuttle Enterprise, covered by an aluminum alloy, made its maiden voyage on top of a specially modified Boeing 747 jetliner, also covered with an aluminum alloy. But in addition to conquering sky and space, aluminum became integrated into daily life through more humble means. In new markets, like beer and soft drink containers, aluminum went from less than two percent of the market in 1964 to 95 percent in 1986.

Legacy

Figure 8.9 An example of aluminum's value to the modern world: the Space Shuttle Enterprise hitches a ride on a 747. Both are covered in aluminum alloys. [NASA.]

But the question remains, is a commanding market share indication of a troubled industry, or can a company be *too good* at its business? That's why Alcoa's story is important to consider. The bottom line is that Alcoa did popularize the use of aluminum among American consumers, made the company's products cheaper, and contributed to the growth of the industrial economy of the 20th century—particularly in vital sectors like the airline industry. But as the name of Alcoa became synonymous with the production of aluminum, it also became notorious among antitrust circles. In the end, the federal government decided that such a large company was naturally incompatible with their view of modern industrial capitalism. The breakup of Alcoa hardly destroyed the company, but it did send a larger message out to firms like Alcoa that might have forged ahead with the large-scale production of their materials. In the postwar American economy, companies still grew large, but often did so with a completely different strategy than Alcoa had employed in its first half-century of growth. So not only did the emergence of Alcoa help structure the American economy of the early 20th century—quite literally in the case of aluminum construction products—but its breakup helped structure the ways in which businesses grew in the late 20th century in the United States.

Discussion Questions

1. Can you think of any other materials that are currently considered luxury products, but might revolutionize everyday life if they became more widely available?
2. Do you agree with Alcoa's early strategy of engaging both in the mass production of aluminum **and** developing new markets for aluminum products? Were company officials seeding their own destruction by trying to do everything with aluminum?
3. Do you think that antitrust actions against Alcoa helped or hindered the American economy in the long run?
4. Do you think that there is a need for antitrust legislation today? Does the federal government need to break up large companies in the interest of competitiveness?

Key Terms

antitrust
monopoly
economies of scale
throughput
horizontal integration
Sherman Antitrust Act
Hall-Héroult Process

Further Reading

Carr, Charles. ALCOA: *An American Enterprise*. New York: Rinehart & Company, 1952. http://www.worldcat.org/oclc/236816.

Peck, Merton J., ed. *The World Aluminum Industry in a Changing Energy Era*. Washington, DC: Resources for the Future, 1988. http://www.worldcat.org/oclc/756989950.

Stuckey, John. *Vertical Integration and Joint Ventures in the Aluminum Industry.* Cambridge, MA: Harvard Univ. Press, 1983. http://www.worldcat.org/oclc/924980139.

Author Biography

Sean Adams is the Hyatt and Cici Brown Professor of History at the University of Florida in Gainesville, where he teaches courses in the history of American capitalism, the global history of energy and 19th century U.S. history. His most recent book on energy transitions and home heating is entitled *Home Fires: How Americans Kept Warm in the 19th Century* (Johns Hopkins, 2014) and it explores the roots of America's fossil fuel dependency during the Industrial Revolution. He is also the author of *Old Dominion, Industrial Commonwealth: Coal, Politics, and Economy in Antebellum America* and a three-volume anthology entitled *The American Coal Industry*, 1789–1902, along with numerous articles, reviews, and book chapters.

Notes

1. George J. Binczewski, "The Point of a Monument: A History of the Aluminum Cap of the Washington Monument," JOM 7 (1995): 20–25, http://www.tms.org/pubs/journals/jom/9511/binczewski-9511.html.
2. Naomi Lameroux, *The Great Merger Movement in American Business*, 1895–1904 (New York: Cambridge Univ. Press, 1988), 164–66, http://www.worldcat.org/oclc/1157046408.
3. "Hall Process: Production and Commercialization of Aluminum," National Historic Chemical Landmarks, American Chemical Society, http://www.acs.org/content/acs/en/education/whatischemistry/landmarks/aluminumprocess.html.
4. Alfred Chandler, *Scale and Scope: The Dynamics of Industrial Capitalism* (Cambridge, MA: Harvard Univ. Press, 2009), 122–24, http://www.worldcat.org/oclc/1041153323.
5. Mira Wilkins, *The History of Foreign Investment in the United States*, 1914–1945 (Cambridge, MA: Harvard Univ. Press, 2009), 32–33, http://www.worldcat.org/oclc/1041152151.
6. George David Smith, *From Monopoly to Competition: The Transformations of Alcoa*, 1888–1986 (New York: Cambridge Univ. Press, 1988), 112–13, http://www.worldcat.org/oclc/908388984.
7. Roland Marchand, *Advertising the American Dream: Making Way for Modernity*, 1920–1940 (Berkeley: Univ. of California Press, 1985), 262, http://www.worldcat.org/oclc/35117815.
8. Smith, *From Monopoly to Competition*, 191–202.
9. Smith, 242.
10. A. Tony Freyer, *Antitrust and Global Capitalism*, 1930–2004 (New York: Cambridge Univ. Press, 2006), 32–40, http://www.worldcat.org/oclc/286424757.

Polymers: Fantastic Plastics in Postwar America

MARSHA BRYANT

> "Plastic is, all told, a spectacle to be deciphered: the very spectacle of its end-products. At the sight of each terminal form (suitcase, brush, car-body, toy, fabric, tube, basin or paper), the mind does not cease from considering the original matter as an enigma." —Roland Barthes[1]

Abstract

This chapter explores how social and cultural systems such as language, gender, aesthetics, home design, and advertising shape the ways we perceive the intrinsic physical properties of materials. Because of their ubiquity in consumer culture, plastics prove especially interesting in these contexts. Even the word *plastic* bears multiple meanings that shape our complicated relationship with this versatile material. The story of Tupperware's invention, distribution, and marketing offers a case study of how materials acquire meanings that shape the ways we publicize and use them. A new polyethylene product in postwar America, the Tupperware bowl became an enduring museum object as well as a household icon.

Introduction

What's in a name? Shakespeare asked in *Romeo and Juliet*. With **plastics,** a synthetic material made from **polymers**, the answer is complicated. Wherever you're reading this, you're likely within reaching distance of at least one plastic product: a cellphone or laptop case, a water bottle, a credit card, a pen, a trash receptacle. Plastics are all around us, every day. Our relationships with plastics can be as richly diverse as the shapes and colors these malleable materials can assume. Plastics also shape our material relations through brand names such as Fiberfil, Mylar, Plexiglas, Teflon, and Tupperware. Even the word *plastic* comes preloaded with contradictory meanings.

Plastics and the Power of Naming

Untangling these scientific and cultural meanings reveals our vexed relationship with plastics. The word's Greek origins link *plastic* to *plassein* (to mold), a dynamic process that makes plastics attractive to manufacturers. Easily shaped through heat and other applications of force, plastics are pliable, ductile, flexible. This technical meaning highlights the material's incredible capacity to transform. In the biological sciences, *plastic* means adaptable to environmental changes. *Plastic* also has aesthetic meanings: three-dimensional modeling, or giving the effect of three dimensions. A more *plastic* model or actor offers more creative possibilities for the canvas or camera. The word can mean an expanded scope for creativity; a medium such as paint or stone can become *plastic* in an artist's hands. All of these positive meanings make plastic products attractive to manufacturers and advertisers.

Figure 9.1 Young Poly Styrene in London. [Photo by Gus Stewart, reprinted in Rolling Stone, June 2012.]

Yet there's a flip side to plastics embedded the same word: *plassein* also means to feign. In social relations, *plastic* can mean superficial or false. A *plastic* performer is someone you'd like to see on stage. But a *plastic* person is someone you'd likely consider phony and avoid. These negative social meanings also affect the material's marketability, even as plastic products continue to fill stores and household shelves. A manufacturer may like plastics because they are cheap to produce. But a consumer may reject a product that looks cheap or gaudy. No wonder we have a love-hate relationship with this material! British punk artist Poly Styrene distilled the tension of plastic's cultural meanings in her stage name. In the late 1970s the punk aesthetic prompted disaffected youth to thumb their studded noses at a British establishment that offered them "no future," as the Sex Pistols proclaimed in their punk anthem "God Save the Queen." Poly Styrene created a rebellious identity for young women through her unconventional attire and anti-commercial lyrics, sporting vinyl accessories and shout-singing edgy songs like "Plastic Bag" with her band, X-Ray Spex. In riotous fashion, Poly Styrene embodied *and* rejected the artificial culture she saw all around her. She took the name of a common plastic and made it a punk icon.

Seeing Plastics through New Eyes

Responding to postwar mass culture in the late 1950s, French literary and cultural critic Roland Barthes included an essay on plastics in his influential book *Mythologies* (1958).

Originally published in a literary journal, the book's analyses ventured boldly beyond Barthes's traditional academic field. He assessed how household products and new technologies transformed everyday life in ways he considered *mythic*: they gave contemporary society its primary means of cultural coherence, functioning as a secular religion. Barthes saw in modern plastics the fulfillment of an ancient dream: *alchemy*, the transmutation of matter. Enchanted by the names of consumer plastics, he imagined **polystyrene**, **polyvinyl**, and **polyethylene** as Greek shepherds in a world of gods and monsters. Instead of viewing plastics through the lenses of science, engineering, and technology, Barthes viewed them through language, literature, and philosophy. The material was magical for Barthes because it was alive with possibility: "More than a substance, plastic is the very idea of its infinite transformation."[2] Sounding as much like an ad man as a philosopher, Barthes pitched plastics as nothing short of miraculous.

How can we see this material through new eyes? How has plastic been a *fantastic* material for inventors, design critics, and advertisers? If we stretch our thinking across academic fields, we can draw from key historical encounters with modern plastics and imagine new possibilities. Earl Tupper, inventor of Tupperware products, saw such promise in postwar polyethylene that he called it *Poly-T: Material of the Future*. At the corporation's Florida headquarters, legendary businesswoman Brownie Wise put raw polyethylene pellets in "Poly Pond"—sprinkling its water on successful saleswomen to give them "Tupper Magic."[3] Perceiving polyethylene with as much imagination as practicality, Tupper and Wise translated their visions into patent applications and successful sales strategies, respectively.

For plastics to fulfill their potential, Robert Friedel wrote in *Pioneer Plastic* (1983), there should ultimately be a "function for every plastic and a plastic for every function."[4] Given plastic's many types and possibilities, this material will be with us for a long time; polyethylene alone reached a global supply of 100 metric tons in 2014.[5] Like other **thermoplastics,** it can assume new shapes and consistencies when heated, melted, and injected into a mold. The polyethylene that Tupper encountered as a plastics sample maker was an industrial waste product—opaque, greasy, clumpy black slag. With these unappealing properties, polyethylene seemed unsuitable for the household market. Tupper would refine polyethylene into a translucent plastic that could take on a range of colors, yielding what journalist Bob Kealing calls "a

Figure 9.2 Vent 'N Serve containers. [Photo courtesy Tupperware Brands Corporation.]

polished, waxy, upscale plastic."[6] You might say Tupper transformed polyethylene from *slag* to *swag*.

How could something as simple as polyethylene become so compelling at the dawn of the space age? How did this material meet people's needs, and how did it change society? Why did design experts love plastics? Cultural historian Alison Clarke explains that Tupperware products *mattered* because they were "at once mundane and extraordinary."[7] As we shall see, Tupper's ingenious refinement of polyethylene and Wise's sales acumen made productive use of plastic's versatility.

How the Tupperware Brand Came to Be

Three key moments in American history shaped the fortunes of Tupperware products' inventor, material (polyethylene), and master promoter: the Great Depression (1929–1940), World War II (1939–1945), and the postwar boom (1946–1960). The first gave Earl Tupper his deeply ingrained values of thrift, resourcefulness, and positive thinking. The second gave polyethylene (and other plastics) new applications in the war industry. And the third gave Brownie Wise an opportune moment to revolutionize direct sales in American homes.

The Homemade Inventor, Earl Tupper

Earl Silas Tupper came of age during the Great Depression, sustained and seasoned by kinship, manual labor, and relentless self-improvement. Born in New Hampshire in 1907 to a farming family, Tupper grew up in a Massachusetts household that often struggled to make ends meet. Ambitious and enterprising since boyhood, Tupper tinkered with ways to improve farm chores. At age 10 he took his parents' produce door-to-door to boost sales. Tupper finished his formal education with high school in 1925, working at the small plant nursery his parents had started after they abandoned farming. He also started his first business as a tree surgeon and landscaper. A self-taught inventor and domestic entrepreneur, Tupper fashioned a wide-ranging personal curriculum that included taking correspondence classes, visiting libraries and trade fairs, reading widely, and writing regularly. His sketchbook and invention journal had plans ranging from egg peeling clamps to a giant theme park he named Cosmopolita World.[8]

While he kept his eye on the future, Tupper drew inspiration from legendary thinkers such as Leonardo Da Vinci, Thomas Jefferson, and Thomas Edison. In his notebook, he wrote: "An inventive, intuitive, enthusiastic and trained mind continues to experiment, study, search and develop along an entire line of human endeavor."[9] Tupper drew from humanities, science, technology, and business to make history as a plastics inventor. We can see his continuing influence on American innovation at the Smithsonian Institution.

The National Museum of American History houses his papers, and the Smithsonian Tropical Research Institute in Panama houses the Earl S. Tupper Research and Conference Center to foster research and development across the sciences.

Earl Tupper's Homemade Curriculum for Innovation

- Technology
- Literature
- Popular magazines (including *American Home*, *Popular Mechanics*, *Literary Guild*)
- Writing (personal journals and notebooks)
- Advertising (correspondence school)
- Industry experience (Doyle Factory, DuPont)
- Following business trends (trade shows, the New York World's Fair)
- Consulting with women about product designs (family, neighbors, coworkers)

The Great Depression and Tupper's farming background played formative roles in how he conceived his breakthrough invention. In an economy of scarcity, thrift and wastefulness loom large; survival depends on making the best uses of what you have. Observing the women working all around him, the young Tupper resolved to make domestic life easier—from cleaning chickens to washing dishes. As Clarke explains, he believed that thrifty provisioning and modern technology could sustain home and society; by contrast, extravagant consumption wasted time and materials. This ideology shaped his early promotions about Tupperware products paying for themselves.[10] Though not cheaply priced, the product's durability and superior seal preserved pantry staples and saved leftovers, stretching household budgets and reducing waste. (The Tupperware Ladies pitched this feature with a new word: *plan-overs*.)

Material Transformations

Figure 9.3 Earl Tupper. [Tupperware.com.]

The plastic material that Tupper would transform for the household market—polyethylene slag—was an industrial waste product he encountered after his tree surgery business failed. In 1937 he moved with his wife and sons to try his fortunes at the Doyle Works, a plastics factory affiliated with DuPont. Through such small operations, DuPont employed amateur sample makers to further plastics research and development. This mutually advantageous arrangement gave Tupper access to scrap material, allowing him to take it home and invent prototypes. When polyethylene production took off in the United States in 1943, the material proved primarily important for wartime uses such as insulation, container linings, cable coatings, gaskets, and tubing. Tupper would *domesticate* polyethylene, viewing the material as a World War II veteran now ready for civilian tasks.[11]

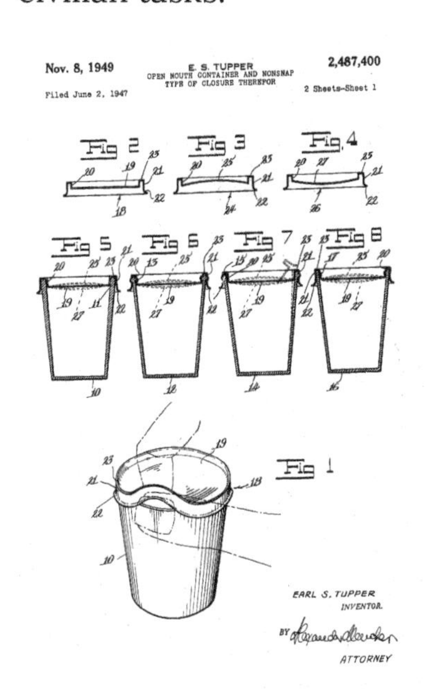

Figure 9.4 Tupper's patent drawings for non-snap lid. [Google Patents.]

Working to overcome the material's limitations, Tupper wanted to produce a plastic more durable than molded transparent styrene. He wanted something that could flex without breaking, something that could stand up to lemon juice and vinegar, something physically appealing to homemakers.[12] In fact, women were integral to Tupper's design breakthrough. He often interviewed female relatives, neighbors, and coworkers to foster ideas for his inventions: a dish rack with drains, better knitting needles, eyebrow dye shields. He had a capable and dedicated partner in his wife, Marie. She would give him tools to improve his sample making, and he would give her his new inventions.[13]

After Tupper's initial failures with the injection molding machines, he took samples home and asked his son Myles to put them in boiling water and remove each at a specified time. Eventually Tupper found the right balance of pressure and temperature, so the polyethylene flowed into various shapes of the desired thickness. He also fashioned a system for dying his translucent containers in pastel colors. But he needed something else: the right lid. As Kealing points out, many American women had relied on tin foil or shower caps to cover their leftovers. Tupper turned to an everyday object for inspiration: paint cans. He fashioned a flexible polyethylene lid that allowed an airtight seal, protected and preserved food, prevented spills. With a simple hand movement, you could manipulate the seal to expel air. He founded the Earl S. Tupper Company in 1939, which became the Tupper Corporation in 1947. That same year, an article in *Time* magazine hailed the inventor for "a process which overcomes the material's tendency to split, makes it tough enough to withstand almost anything except knife cuts and near-boiling water."[14] The toughness of Tupperware products makes a comic appearance in the film *Napoleon Dynamite* (2004) when salesman Uncle Rico urges a client to tear the product with his bare hands—an impossible task.

Activity: Explore the Museum of Modern Art Website

Navigate to the MoMA Learning website and search for "tupperware." How does the product's display as a museum object change your perception of its material?

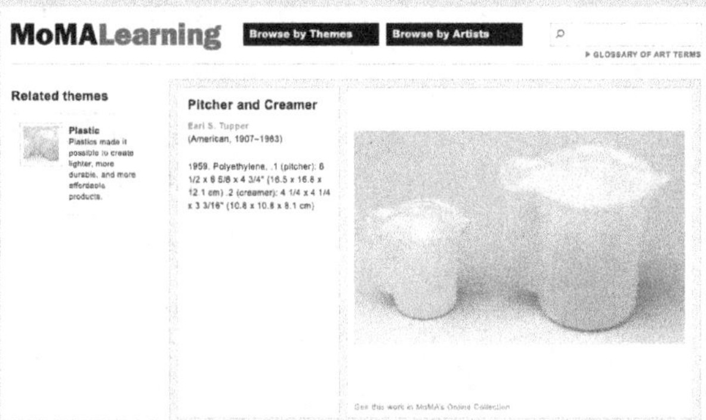

MoMA *Learning*

Click *Browse by Theme* and enter *plastic*. Click the Theme box *Plastic* at the top left, and check out the resources. If you click *Tools & Tips*, you'll find worksheets on design and environmental issues.

On the same page under *Multimedia*, you'll find a brief video on plastics produced in 1944 for the Young America series. How would you promote plastics to your own generation?

Polyethylene as Everyday Art Object

In 1948, the Tupper Corporation's tea set made the *Modern Plastics Encyclopedia* as an ideal incarnation of domestic polyethylene. If you think about it, plastic seems an unlikely material for this purpose. We don't usually count plastics with traditional wedding gifts and family treasures: silver, china, and crystal. Yet we find the Tupperware pitcher and creamer in the Museum of Modern Art (MoMA), along with Tupper's tumblers, bowls, and "Ice-Tups" (his ingenious popsicle molds). MoMA curators included Tupperware products in "Good Design" exhibitions in the 1950s, as well as the 2011 exhibition "What was Good Design? MoMA's Message, 1944–1956."

Key Concept: Modernist Aesthetics 101

"No ideas but in things." —William Carlos Williams, physician and poet

- Form over function
- Autonomy of objects
- Objects can transcend the ordinary
- Clean lines, geometric designs
- Classical restraint over Romantic excess
- Authenticity
- Directness
- Experimental form

"Not Ideas about the Thing, but the Thing Itself." —Wallace Stevens, business executive and poet

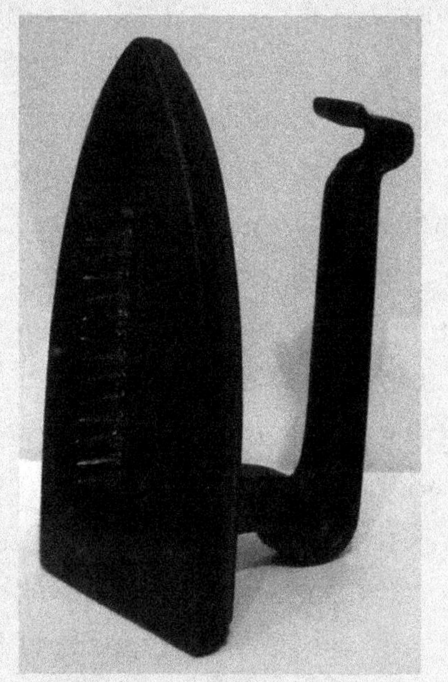

"Cadeau" (The Gift). [Sculpture by Man Ray (1921), Tate Modern. Wikipedia Commons.]

Just as Tupper made plastic respectable for middle-class Americans, he turned polyethylene into an artistic medium. In 1947, an editor for *House Beautiful* enthused that Tupperware bowls "have a profile as good as a piece of sculpture," declaring that the product satisfies "our *aesthetic* craving to handle, feel and study beautiful things." Tupper's timing was good because his product fit the aesthetics of **modernism**, an established style in museums, publishing, and universities. Tupperware products redeemed common plastic through a superior *form* that fused beauty with utility. As Clarke explains, they fulfilled modernism's "ideal of a tasteful, restrained, and mass-produced artifact, free of inauthentic decoration and gratuitous ornament."[15] The product brought class to "mass." And above all, it was *true to its material*. You can see the modernist heritage of Tupperware products in this contemporary video ad for the Tupperware Breakfast Maker. Note how the camera highlights the sleek, colorful design, bringing plastic to life.

How Tupperware Became an American Icon

Tupper's timing was also fortuitous because of postwar changes in American home and family life—the third historical context that shaped his product's fortunes. New social spaces—suburbs—refashioned family relations for many Americans, prompting them to move away from the city (and their extended families). As historian Elaine Tyler May points out, "the legendary white middle-class family of the 1950s, located in the suburbs, complete with appliances, station wagons, backyard barbecues, and tricycles scattered on the sidewalks, represented something new."[16] These postwar couples in their ranch houses were suburban pioneers. Like their mythic counterparts Ward and June Cleaver on television, the Breadwinner and Housewife roles created a gendered division of labor outside and inside the home. Breadwinner donned his gray flannel suit for his morning commute to the office, conforming to a role that social analyst William Whyte termed the Organization Man. Housewife provisioned her home and performed the three C's: cleaning, cooking, and childrearing. She conformed to the new standard of professionalized homemaking, and she was an *ideal* customer for Tupper's ingenious products. May points out that housewives wielded considerable buying power in postwar America: "In the five years after World War II, consumer spending increased 60%, but the amount spent on household furnishings and appliances rose 240%."[17] But even so, Tupperware products weren't flying off the shelves.

The Savvy Saleswoman, Brownie Wise

Brownie Wise changed Tupper's fortunes when he noticed her unusually large orders coming out of Detroit in 1949. A door-to-door seller for Stanley Home Products, Wise had formed her own side business to promote Tupperware products with "Poly-T Parties." She found the Wonder Bowl distinctive enough to sell, but she fumbled with the seal—until she realized that "you had to burp it just like a baby."[18] And with that personal analogy Brownie Wise made Earl Tupper's product come alive in American homes. A television commercial from the 1950s shows a Tupperware party; note the female voiceover and the hands-on demonstrations.

Wise did not see Tupperware products as museum objects to be looked at; they were beautiful things to touch and handle. Her early sales plan insisted that a good "visual demonstration" is the "means of creating a desire for the product."[19] Wise knew how to handle her material, too. In her hands, Tupperware products burped up sales. In her hands, Liquid-filled Wonder Bowls flew across American living rooms, astonishing housewives with that airtight seal. In her hands, the Tupperware Home Parties Division made Tupperware a household word. Tupper made her general sales manager in 1951, and she became a vice president.

Figure 9.5 *Brownie Wise.*
[*Business Week* (April 17, 1954).]

The first woman to appear on the cover of *Business Week* (April 1954), Wise transformed materials marketing as dramatically as Tupper had transformed polyethylene. With even less formal education than Tupper, Wise overcame financial hardship with resourcefulness and determination. A divorced single mother in Detroit, she turned to sales to supplement her low-wage secretarial job and pay her son's medical bills. Her sales experience helped Wise hone her techniques for selling Tupperware products and recruiting dealers. As Kealing explains, she turned away from the "hard sell" for a softer approach that stressed "the social and emotional aspects" of the transaction.[20] Wise praised her Detroit dealers for stand-out parties and sales figures in her upbeat newsletter, the *Go-Getter*; it eventually became the *Tupperware Sparks* publication for a national sales force. Wise wrote guidelines for successful demonstrations—and for sizing up potential Tupperware Ladies. She also published her autobiography, *Best Wishes, Brownie Wise* (1957), reprinted 40 years later to motivate a global sales force.

Wise convinced Tupper to fundamentally change his marketing strategy. Taking his product off the shelves and out of the stores, her business plan put more Tupperware products into more women's hands. Her Tupperware Home Parties Division moved to Florida in 1950. And the woman once told that women couldn't advance at Stanley became a business celebrity and inspirational figure. At her Jubilee conventions in Orlando, Wise offered Tupperware Ladies fur coats, jewelry and other fine things as sales incentives. You can see Wise work her sales magic in Laurie Kahn-Levitt's film Tupperware! (2004), aired on PBS. Though the product bore Tupper's name, Brownie Wise became the face of the company—a factor in Tupper's decision to fire her in 1958. But Wise proved pivotal to making Tupperware Brands Corporation the successful enterprise it remains today. She also gave postwar American housewives paid work within the home that was socially acceptable. Part housewife, part hostess, part businesswoman, the Tupperware Lady straddled the era's social categories, crossing geographic, racial, and class boundaries.

There were also husband-and-wife teams who worked as Tupperware dealers inside and outside the home.

American Advertising and Materials Marketing

The postwar boom and dream kitchen ushered in new ways of marketing household products as advertisers courted American housewives. Like *Mad Men*'s Don Draper, they competed relentlessly for women's attention. How could they make their client's product stand out from the other brands? Tupper had taken an advertising correspondence course, and he kept in touch with his ad executives. He wrote to one that "a product is not exactly what you make it in the factory, but it is also and more so what you make it when you sell it."[21] A successful advertisement brings together the manufacturer, product, and consumer in a dynamic relationship. Revisiting milestones in American advertising can help us understand the many ways consumers connect with products and their materials. Juliann Sivulka explains that postwar advertisers began to move away from hard-selling product features, drawing from psychology "to build a product image or brand personality."[22] The three classic campaigns I discuss here take different proximities to the product, emphasizing in turn the consumer, the product itself, and the desired image.

Listerine and Selling the Need

The first shows Gerard Lambert's strategy of *selling the need*. His Listerine campaign made the medical term for bad breath—*halitosis*—a household word. It all started in the late 19th century with a surgical antiseptic that physician Joseph Lister formulated in England. In 1879, Joseph Lawrence made a milder version in the United States, naming it after Lister and promoting it as an all-purpose antiseptic. First successful with dentists, Listerine entered the mass market in the 1920s. By then the Lambert Pharmaceutical Company had acquired the product, and Gerard Lambert was seeking greater profits. Learning about halitosis from a company chemist, Lambert seized upon the word as a way to get Listerine into more mouths.[23] Halitosis *sounded* as unseemly as it was. Making people aware of it would make them want to get rid of it.

We can see how the pitch worked in this top half of a 1928 Listerine ad. If you take a close look, you'll see how efficiently the ad's words and image distill its claims and implications:

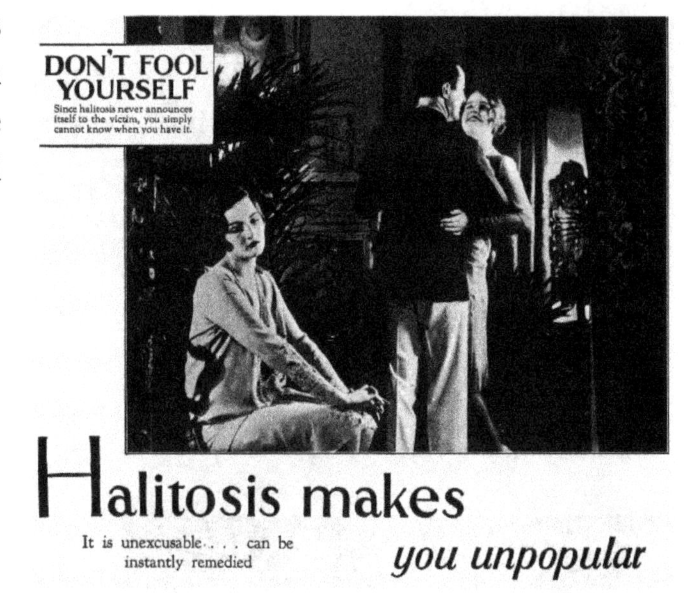

Figure 9.6 Excerpt from 1928 Listerine ad. [Reprinted in Smithsonian Magazine, 2015.]

- The textbox at the top left makes *no one* immune. Halitosis strikes at any moment!
- In the image, décor and attire show that even respectable people suffer from halitosis.
- The afflicted woman in the corner stands out by sitting out this dance. She looks exposed and uncomfortable, unlike the smiling woman dancing with her partner.
- By implication, halitosis kills any chance at romance.
- The copy beneath the image makes *us* the culprit for halitosis. With Listerine so readily available, bad breath becomes an unforgivable faux pas.

In his bestselling book *Advertising the American Dream* (1985), historian Roland Marchand explains that such ads play on "the Parable of the First Impression"—a powerful theme in ads from the early 20th century.[24] (210). Selling the need to avoid embarrassment through the invisible menace of halitosis, Lambert's Listerine campaign made antiseptic mouthwash the antidote to unpopularity.

Anacin and the U.S.P.

The second advertising milestone powered Rosser Reeves's Hard Sell technique—the *Unique Selling Proposition* (U.S.P.). A contemporary of Tupper's, Reeves did his most influential work with the Ted Bates ad agency in New York in the 1940s and 1950s. (He had first worked as a journalist.) In his bestselling book *Reality in Advertising* (1961), Reeves distinguished the U.S.P. from other strategies as an appeal to the consumer's *reason.* He believed that a successful ad is efficient; it zeroes in on a precise claim about the product and avoids any distractions. Moreover, a successful ad avoids puffery and fantasy, grounding appeals in the product and reality.

We can see these strategies at work in Reeves's famous Anacin commercial that featured an animation of a headache; look for it 35 seconds into this video from 1958.

Invoking the medical profession to enhance credibility, this Anacin ad zooms in on the headache and brings it to life. Like the pounding hammer in the middle panel, Reeves's ad hits the reader in the head with his U.S.P.—*fast*. The repetition links the U.S.P. with other things that come in threes: e.g., "three out of four doctors recommend"; Anacin has three pain relievers; it relieves all three headache symptoms. The ad distinguishes its product from competitors' and offers a specific benefit: Anacin will stop your headache fast, Fast, FAST. Sivulka points to contemporaneous U.S.P. ads that also proved effective: *Certs breath mints with a magic drop of Retsyn; Colgate cleans your breath while it cleans your teeth.*[25]

Hathaway and Branding

Figure 9.7 The Man in the Hathaway Shirt.

The third advertising milestone, David Ogilvy's concept of *branding*, emphasizes the consumer's desires as much as the product itself. Reeves saw branding this way: "To put it bluntly, the U.S.P. is the philosophy of a claim, and the brand is the philosophy of a feeling."[26] But there was nothing blunt about Ogilvy's creative approach to advertising:

> Products, like people, have personalities, and they can make or break them in the marketplace. The personality of a product is an amalgam of many things—its name, its packaging, its price, the style of its advertising, and, above all, the nature of the product itself.[27]

Ogilvy took the idea of products and personality beyond personified objects (the "Tupperware burp") and corporate personas (Betty Crocker, Uncle Ben). Ogilvy had sold cooking stoves door-to-door in England before moving to the United States in 1938. A decade later, he started a New York ad agency. Ogilvy became legendary in American advertising through his successful campaigns and his bestselling book *Confessions of an Advertising Man* (1963). He is the historical inspiration for the character Don Draper on *Mad Men*.

Ogilvy's campaign for Hathaway also became legend. Drawing on the principle that consumers are less interested in the product per se than the *image* they buy with it, this ad creates a compelling character: "the man in the Hathaway shirt." No conformist

Organization Man in a gray flannel suit, this man was singular and sophisticated. He wore an eye patch. He could sail, paint, fence, ride horses, and play the oboe. A model of cultural refinement, he traveled the world. The man in the Hathaway shirt embodied the brand's new personality. Launched in 1951, this campaign initially ran exclusively in *The New Yorker.*[28]

The Hathaway man's descendants appear in contemporary ads for Dos Equis beer (The Most Interesting Man in the World) and Old Spice deodorant (The Old Spice Guy who smells like a man). But advertising was never strictly a man's world. Ogilvy's campaign succeeded in part because its international man of mystery appealed to *women* as well as men; after all, women were the primary shoppers in postwar America. Sivulka and other cultural historians have now turned their attention to women who work in advertising. Where would Don Draper be without Peggy Olson?

Conclusions: Tupperware and the Future of Plastics

Revisiting vintage promotional materials for Tupperware products reveals social and cultural meanings this iconic plastic conveyed to middle-class housewives in the 1950s–1970s. It also prompts us to think about how plastic's properties might appeal to specific demographics in contemporary society. In this catalog cover from 1958, for example, Joe Steinmetz's photograph places the woman in two key contexts: the family circle and the museum gallery. In the first, her provisioning enables the genial gathering we see in the background. Framed between glass shelves, just right of center, her husband smiles as he holds his Tupperware mug (Tupperware products also appear on the table). Mother is the central figure, arranging her Tupperware collection in a beautiful display that resembles a museum's—the second context. She smiles at her Tupperware "family" as she would at the family seated behind her. Sleek, colorful, and modern, her Tupperware products have brought harmony to her kitchen and home. Everything is orderly; nothing is out of place.

Figure. 9.8 Catalog cover. [Photo by Joe Steinmetz (1958). Florida Memory.]

The American postwar dream kitchen gleamed with shiny metal appliances, polished wood cabinets, and colorful linoleum floors. But plastics also transformed this vital space. A fantastic polymer, plastics can take on pretty much any shape and color we can imagine. Brownie Wise marveled at "the tremendous imagination it took" for Earl Tupper to change black polyethylene slag into his signature products.[29] More recently, Tupperware Brands Corporation has designed UltraPro Ovenware products that "can be used in temperatures up to 482° F/250° C and as low as -13° F/-25° C."[30] The new copolymer ethylene-tetra-fluoro-ethylene (ETFE) has transformed the roof of the Minnesota Vikings' football stadium, letting in light with lightweight transparency—and costing less than glass. This material also formed the dramatic bubble façade of the Water Cube venue for the 2008 Beijing Olympics. Where will plastics go next? Your generation will create imaginative designs for improved and future products, inventing new brand names for this fantastically versatile material.

Tupperware Milestones in Popular Culture

- 1974. *All in the Family* television series. Edith Bunker hosts her first Tupperware Party. (Season 5, episode 8)
- 1986. Tupper Ware Remix Party, a futuristic synthwave dance band, forms in Toronto, Canada.
- 1995. *Seinfeld* television series. Kramer tries to retrieve his Tupperware container from a homeless man. (Season 6, episode 15).
- 2004. *Napoleon Dynamite* film. Uncle Rico tests Tupperware's toughness with a client, and Kip tests it with a van.
- 2015. *American Horror Story* television series airs "The Tupperware Massacre" episode. (Season 4, episode 9).
- 2015: Hollywood film about Brownie Wise goes into production

Discussion Questions

1. Get your hands on a piece of Tupperware. What properties do you notice?
2. Before Tupperware hit the market, most Americans stored their food in glass. Why is polyethylene a better material for this purpose?
3. Which household appliance do you think most shaped the fortunes of Tupperware? Why?
4. In postwar America, common thinking held that married women should spend most of their time in the home—especially if they were mothers. How did Brownie Wise make it socially acceptable for such women to sell Tupperware?
5. Watch this video "Tupperware Celebrates World Water Day 2013." What are the environmental impacts of plastic products? How might they contribute to sustainability?
6. If you were to put your name on a plastic product, what would it be? Which kind of plastic do you think would be ideal for your purpose?

Key Terms

modernism
plastics
polyethylene
polymers
polystyrene
polyvinyl
thermoplastics

Author Biography

Marsha Bryant researches and writes about modernist studies, visual culture, women's writing, popular culture, and pedagogy. She is Professor of English & Distinguished Teaching Scholar at the University of Florida, where her most popular courses include modern poetry surveys, *Desperate Domesticity: the American 1950s*, and *PostPunk Cultures: The British 1980s*. Her recent essays have appeared in *The Massachusetts Review* (about Walt Whitman and beer culture), *Feminist Modernist Studies* (about Sylvia Plath and Edith Sitwell), and *The Conversation* (about Tupperware). Professor Bryant's book *Women's Poetry and Popular Culture* received funding from the National Endowment for the Humanities. She is a member of UF's Academy of Distinguished Teaching Scholars and a UF Doctoral Mentor awardee. Professor Bryant is also Associate Editor of *Contemporary Women's Writing*.

Notes

1. Roland Barthes, *Mythologies*, trans. Annette Lavers (New York: Hill and Wang, 1987), 87, http://www.worldcat.org/oclc/18888151.
2. Barthes, *Mythologies*, 97.
3. Alison J. Clarke, *Tupperware: The Promise of Plastic in 1950s America* (Washington: Smithsonian Institution Press, 1999), 137, http://www.worldcat.org/oclc/634385344.
4. Robert Friedel, *Pioneer Plastic: The Making and Selling of Celluloid* (Madison: Univ. of Wisconsin Press, 1983), 108–109, http://www.worldcat.org/oclc/9081371.
5. Vincent Valk, "Outlook 2014: Looking Forward," *IHS Chemical Week*, April 14, 2014, http://www.chemweek.com/lab/Outlook-2014-Looking-forward_57898.html.
6. Bob Kealing, *Tupperware Unsealed: Brownie Wise, Earl Tupper, and the Home Party Pioneers* (Gainesville: Univ. Press of Florida, 2008), 22, http://www.worldcat.org/oclc/185123372.
7. Clarke, *Tupperware*, 4.
8. Clarke, *Tupperware*, 8–12, 15–17, 19–20; "Earl Silas Tupper," American Experience, PBS, http://www.pbs.org/wgbh/americanexperience/features/biography/tupperware-tupper/.
9. Qtd. in Kealing, *Tupperware Unsealed*, 10.
10. Clarke, *Tupperware*, 10, 64, 114.
11. Clarke, 26–29, 37–38.
12. Kealing, *Tupperware Unsealed*, 21.
13. Clarke, 30–34.
14. Kealing, 21–22.
15. Clarke, 42, 49.
16. Elaine Tyler May, *Homeward Bound: American Families in the Cold War Era*, 20th Anniversary Ed. (New York: Basic Books, 2008), 13, http://www.worldcat.org/oclc/690487945.
17. May, 207.
18. Qtd. in Kealing, *Tupperware Unsealed*, 26.
19. Qtd. in Kealing, 38.
20. Kealing, 44.

21. Qtd. in Kealing, 23.
22. Juliann Sivulka, *Soap, Sex, and Cigarettes: A Cultural History of American Advertising*, 2nd ed. (Boston: Wadsworth, 2012), 231, http://www.worldcat.org/oclc/714878884.
23. Jesse Hicks, "Thanks to Chemistry, A Fresh Breath," Chemical Heritage Foundation, https://web.archive.org/web/20160407023152/http://www.chemheritage.org:80/discover/online-resources/thanks-to-chemistry/listerine.aspx
24. Roland Marchand, *Advertising the American Dream: Making Way for Modernity*, 1930–1940 (Berkeley, CA: University of California Press, 1985), 210, http://www.worldcat.org/oclc/35117815.
25. Sivulka, *Soap, Sex, and Cigarettes*, 232.
26. Rosser Reeves, *Reality in Advertising* (New York: Alfred A. Knopf, 1968), 79, http://www.worldcat.org/oclc/7540914.
27. David Ogilvy, *Ogilvy on Advertising*, 1983 (New York: Vintage, 1985), 14, http://www.worldcat.org/oclc/613287090.
28. Sivulka, *Soap, Sex, and Cigarettes*, 236.
29. Qtd. in Kealing, *Tupperware Unsealed*, 56.
30. UltraPro Ovenware product insert, Tupperware.

Writing Materials: The Politics and Preservation of Knowledge

BONNIE EFFROS

> "The library of the future must also be a place that will preserve the knowledge and understanding of written culture in the forms that were and still are today, very much its own." —Roger Chartier

Introduction

Just as it is very challenging to communicate to those around us without gestures or spoken language, it is difficult to convey thoughts and desires in a more lasting manner (or to those at a distance) without either the written word or pictorial representation.[1] In Europe as early as the Renaissance, scholars pondered the origins of language as intimately tied to writing symbols. They believed themselves a part of a grand cultural tradition that traced back to classical antiquity, and linked their own achievements with those of the ancient Greeks and Romans.[2]

Giambattista Vico (1668–1744), a humanist philosopher in Italy, elevated the achievements of the ancient Greeks over those of all earlier civilizations because he believed they were the first to master alphabetic writing. In his view, mastery of writing in words as opposed to the hieroglyphic symbols of the Egyptians, freed Greek citizens from having to rely on priests or aristocrats. Rather than a "secret language" like cuneiform or hieroglyphs, what Vico described as **"vulgar writing"** was a key ingredient in the origin of democracy in classical Greece. From his perspective, alphabetic writing stripped figures and signs of their mysteries and gave people equal access to knowledge of truth. Letters made it possible for a greater number of people to learn to read, which, in turn, was intimately connected to their ability to participate in both religious worship and government in a direct and personal manner.[3]

Figure 10.1 In 1799, a French soldier during the Napoleonic invasion rediscovered the Rosetta Stone (196 BCE) in the town of Rashid (Rosetta) in the Nile Delta. When the British navy defeated the French in 1801, they took it to the British Museum in London where it remains today. In 1822, the French scholar, Jean-François Champollion, succeeded in translating the decree issued by Ptolemy V transcribed in three languages on the Rosetta Stone. He was able to decipher the hieroglyphs and demotic Egyptian in the top two registers of the inscription with the help of the Greek in the lowest register. [Wikimedia Commons.]

Although the transition to an alphabetic system was an important step in the development of written language, we now recognize the problematic implications of Vico's Eurocentric perspective: it was based on his implicit assumption that all parts of the world were inferior to those that followed the European tradition. We can alter his observation in several ways to make it more accurate and helpful to our purposes here. First, we can point to other languages that made the transition to alphabetic systems in antiquity, including among them the Ge'ez or Ethiopic syllabic alphabet used from the 4th century BCE, the Brahmi writing system of South Asia seen as early as the 3rd century BCE (a forerunner to Sanskrit), and the Tifinagh or Libyc syllabic alphabet used in the Maghreb (North Africa) from some time in the 2nd millennium BCE. Second, we can suggest that alphabets were never a guarantee that large numbers of people could read them; literacy rates depend far more on access to education than the expressive format of any particular language. We can also look to China, for instance, where millions have mastered the ability to read using the character system despite its evident challenges (yet recognize that this influential system was modified in Japan and later Korea so as to be more accessible). Finally, we can underline the fact that Greek democracy no longer seems as enviable as it was to Vico, since it only allowed elite men (and not women, the non-free, and non-Greeks) to vote (and read) rather than a broader sector of the population. That being said, the development of alphabets indeed made the process of learning to read and write more efficient by decreasing the number of symbols that it was necessary to memorize.

Back in Europe in the decades prior to the French Revolution, Enlightenment thinkers continued to wrestle with the significance of language in social and political development. However, they now placed greater worth on reason and individual thinking than on ancient tradition, as had been the case during the Renaissance. The philosopher Nicolas de Condorcet (1743–1794) pointed to the positive social good brought by recent inventions such as the printing press. Condorcet believed that the limitless reproduction of texts afforded by moveable print was key to the success of democracy, which he experienced firsthand as a delegate to the National Assembly at the start of the French Revolution. In fact, he credited this technological achievement with greater access to justice in human society. In his view, the proliferation of the written word broke the Church's monopoly over knowledge by allowing more individuals access to texts which they could read and judge for themselves.[4]

Those among you who love numbers will be pleased to learn that Condorcet identified mathematics as the most universal and most perfect form of language. He viewed numbers and formulas as a highly developed alphabet and form of communication. Since most women and the poorer classes of his day had insufficient education to participate in this dialogue, Condorcet believed that improved schooling was essential to a just society.

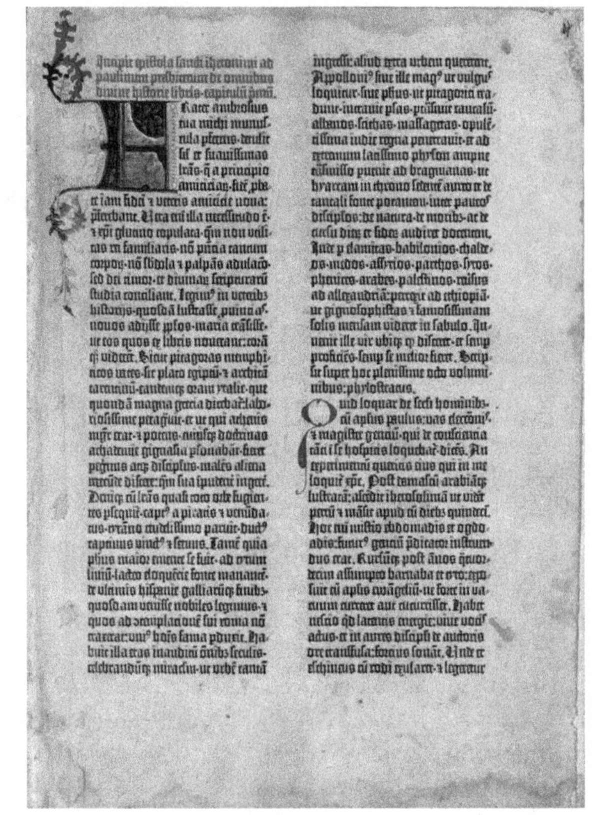

Figure 10.2 This opening page of the Gutenberg Bible at the University of Texas (Hubay 39) contains Saint Jerome's introduction. One of the earliest printed books in Europe, the style imitates a medieval manuscript with historiated initials, illustrations and double columns of Latin text on a single leaf. Unlike a hand-copied manuscript, it could be reproduced multiple times once the type had been set. [Johannes Gutenberg (1454-56), University of Texas Harry Ransom Center.]

These century-old debates about the origins of language—and the direct connection between the written word and participatory democracy—may seem part of the distant past to many of us. Yet, they are relevant because they offer unique perspectives on the challenges and opportunities of the digital revolution and recent innovations in magnetic storage. These early modern examples offer a historical comparison that gives us a basis for measuring the impact of writing materials on society. They allow us to ask if technological innovations like the internet and the ability to store ever-increasing amounts of data are actually making us more knowledgeable and improving the quality of our lives. They also give us the chance to consider whether the cost of entry (access to a computer or lack of education) excludes people who might benefit from this information.

Understanding these discussions by Enlightenment thinkers also helps us to formulate vital questions about who controls knowledge in the digital age and what obstacles exist to full participation. Thankfully no longer restricted just to the educated elites, humanistic thought encourages us to apply greater objectivity to the exploration of the politics of collecting, storing, and transmitting information both in the past and today. Whatever our conclusions, we cannot deny the deep and meaningful connections that exist between writing forms and media (i.e., the materials we use and manipulate), and between the preservation of knowledge and the organization of human society.

Humans and the Search for Functional Writing Materials

Figure 10.3 This Chinese text was copied on bamboo strips using a brush and dates to the Warring States Period (475–221 BCE). These strips, read downwards and from right to left, are now held in the Shanghai Museum, and contain a discussion of the 'Shi Jing', or Book of Odes. [Wikimedia Commons.]

From the earliest times, humans used various media for writing surfaces: the walls of caves, stones, bones, bamboo strips, and ceramic sherds. Once inscribed with markings, these materials transmitted messages of religious, ritual, magical, and practical significance like trade inventories, curse tablets, and to-do lists. However, not all of these materials were equally accessible and efficient. They varied regionally in response to the availability and affordability of particular resources. In addition, written language was often out of reach for the vast majority of pre-modern populations, which were for the most part functionally illiterate.

In ancient Babylonia, priests used a metal stylus to mark soft clay tablets with cuneiform texts when they wanted to preserve their writing for the elite audience who could read them. Cylinder seals, rolled over a similar medium, were particularly handy when priests wanted to reproduce texts multiple times. Once baked, such inscribed clay tablets were sufficiently durable to last millennia under the right conditions.[5]

By contrast, the ancient Greeks and Romans chose stone carving for important messages that they intended to display in posterity such as grave markers. However, in daily life, for the sake of convenience and affordability, they often relied upon handy potsherds or **ostraca**, broken bits of pottery that provided a smooth surface and inexpensive material for writing. Wax-coated tablets in the ancient world, by contrast, offered the benefit of reuse after heating. This medium which was especially suited to classroom use, where children were learning to write. In 8th-century Korea and China, engraved wood allowed repeated reproduction of the written word by rubbing, a technique known as xylography that was transmitted to the West only half a millennium later.[6]

Figure 10.4 This terracotta clay cone inscribed in cuneiform with a stylus commemorates the construction of the walls of Sippar by King Hammurabi. Now at the Louvre Museum, it dates to the first half of the 18th century BCE and comes from Iraq. [Wikimedia Commons.]

Figure 10.5 This late 3rd-century ostracon in Greek preserves a private letter concerning the use of oxen for sowing a field. It suggests that immediate tasks could be written on a variety of accessible surfaces and were not intended by their authors to be kept in perpetuity. ["O.Mich.inv. 4194; Recto," University of Michigan Library, Advanced Papyrological Information System.]

In China, the use of silk as a writing material showed the advantages of shifting to a more flexible writing medium. However, it was very costly and limited to the elites. In the West, the ancient Egyptians were among the first to employ more flexible writing media, namely papyrus. Egyptian artisans manufactured this material from the papyrus reed, which grew mainly in the region known as the Egyptian Delta. To prepare the reeds, Egyptians cut the pithy centers of stalks of the plant into thin strips which they then placed side-by-side with superposed layers at right angles to form sheets. Pressed and beaten together while still moist, the fibers adhered even when dry. The horizontal lines left by the papyrus strips became the guide for scribes to keep their writing (done with a reed brush) straight and level. Papyrus, however, was unforgiving in a number of ways if a scribe made an error. While erasure was possible with a sponge or an abrasive substance, it was not easy to write over the same spot. Another downside to the material was that the sheets became brittle when dry and thus could not be folded. For this reason, papyrus sheets were rolled as a scroll. Likewise, scribes could only write on one side because ink bled through the semi-porous material.[7]

Activity: Browse a slideshow

University of Michigan

This exhibit (https://apps.lib.umich.edu/papyrus_making/#) from the University of Michigan Papyrus Collection demonstrates how papyrus is manufactured, from harvest to finished paper.

Papyrus reeds grew in few places other than Egypt and Ravenna on the Adriatic coast of Italy. Egypt thus exercised a near monopoly on the production and trade of this elegant, light-colored writing substance throughout classical antiquity. It became popular not just among the Egyptians but also among the Greeks and Romans, who were attracted to the relative affordability of papyrus rolls. With the advent and spread of Islam through the Middle East and North Africa in the 7th century, papyrus remained an important writing material for adherents of the new religion. Muslim administrators used the medium for official correspondence, legal documents, ledgers, and tax receipts. In northern Europe, however, climatic conditions supported neither the cultivation of papyrus nor its effective use since humidity hindered its preservation. Few surviving examples of papyrus in Europe date after the 7th century.[8]

Does Form Determine Function?

In antiquity, like today, specific qualities of writing materials dictated the purposes for which they were employed. As we have seen, stone, for instance, was most suitable for epitaphs or public proclamations, whereas contemporaries did not see papyrus as having sufficient permanence and durability for long term preservation. For these reasons, papyrus was not commonly used for longer literary or sacred texts. The accessible and cost effective medium nonetheless remained popular for day-to-day transactions like trade in the dry, hot conditions of North Africa and the Middle East. These examples should make us think about more recent technological developments and to what degree digital media like electronic books and newspapers are better suited to some objectives (like quick reading) than others (note-taking and deep engagement with a text).

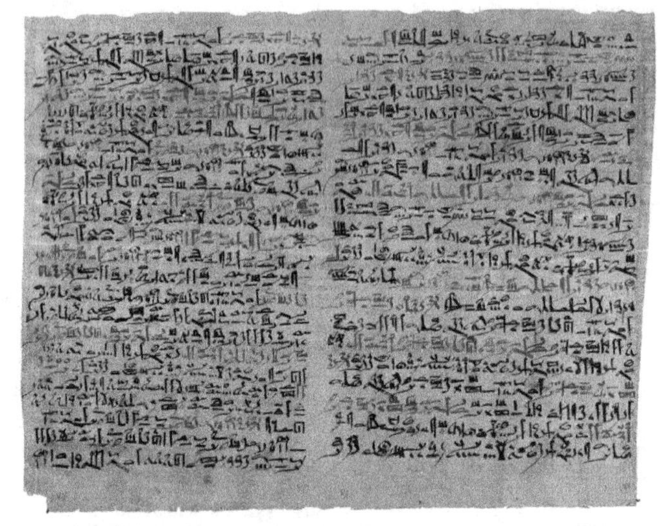

Figure 10.6 Dating to the Second Intermediate Period in Egypt, the Edwin Smith papyrus may have come from the capital of Thebes. Purchased in Luxor, this hieratic manuscript (Egyptian cursive) was donated to the New York Academy of Medicine. It is thought to be the oldest surviving Western text of surgery, and includes material on surgery, gynecology, and magic. [New York Academy of Medicine. Wikimedia Commons.]

Something to keep in mind is that the format of papyrus, namely scrolls, influenced the way in which authors composed texts and people read them. The manufacture of papyrus rolls in standard lengths necessitated the division of literary works into books. Anyone familiar with the ancient Greek epic poems of the *Iliad* and *Odyssey*, for instance, should consider that scholars of Alexandria apportioned this work of the 8th century BCE into books relatively arbitrarily based on how much material fit on a scroll. By the time of the Latin author Virgil in the 1st century BCE, however, authors began to think of book divisions—today's chapters—as discrete parts of a work. In other words, writers took advantage of divisions introduced by their writing medium to structure their texts. In the *Aeneid*, Virgil applied this technique to represent changes in the narrative.[9]

Figure 10.7 This scribe seated with a papyrus scroll on his lap suggests some of the difficulties in using a scroll for study. It is hard to mark one's place! Dated to the 19th–20th dynasty, 1295–1069 BCE in Egypt, this diorite statue gives us some clues as to why a codex was preferable if one wanted to navigate quickly through a text. [Louvre Museum. Photo by user Janmad, shared under a CC BY 3.0 Unported License. Wikimedia Commons.]

Increased familiarity with the medium of the codex—what we know today as a book (see below)—gave birth to an intellectual shift and a revolution in literature which became dominant by the 4th century CE. We see similar steps being taken today, as engineers continue to modify the format of electronic books to meet the needs and expectations of readers. Whereas engineers designed the earliest Kindles to look like books with a plain white background, black print, and limited possibility to alter texts, today's touch screens allow readers freedom to manipulate the content and appearance of their reading.

Putting Quills to Parchment

Contemporary to papyrus, parchment was a writing material made from animal skins. In the ancient Middle East, parchment was commonly made from skins of calves, sheep or goats. White animals were preferred since they tend to produce the lightest colored parchment. Vellum, the most desirable form of parchment because of its refined surface, referred exclusively to calfskin. Parchment, which was far more costly than papyrus, was developed first in ancient Egypt, Assyria, and Babylonia, but then spread to many regions that raised flocks. It provided material support for writing, painting, and other purposes. Despite its durability and versatility as a writing surface, parchment's cost was prohibitive. This technology thus never fully replaced the use of papyrus, ostraca, stone, and wax tablets for recording information.[10]

To prepare animal hides for writing, workers soaked the skins in a lime solution and removed the remaining hair and flesh (a characteristic that survives in parchment manuscripts, in which the hair and flesh sides are discernable). Workers washed and stretched the cleaned hides tightly on a wooden frame and scraped them with moon-shaped knives as they dried. When the hides were nearly dry, workers buffed the surfaces with pumice stone. From the 4th century CE, scribes began to fashion split quills from feathers as a firmer tool for writing than reed brushes. Quills quickly supplanted reed brushes for writing in the West due to their availability in virtually any market. They offered exciting possibilities for artists who learned to incorporate shading and decorative work in written characters that had not been possible with reed brushes. The spread of parchment and quills contributed to the development of new letter forms and book design features.

Parchment offered the advantage of lightness and flexibility, which meant ease of handling and transmission. The surfaces were suitable for inks and colors and one could write on both sides. Scribes could make erasures and corrections. Parchment also accommodated folding and stitching and could be sewn together in scrolls or organized concertina fashion since sheets of parchment did not tear easily as papyrus. With a long life, parchment could be stored for more than a millennium in the right conditions and wear was minimal compared to papyrus. Due to cost, however, access to parchment was limited to those of financial means such as religious institutions and individuals of high status such as pharaohs, kings, and aristocrats.

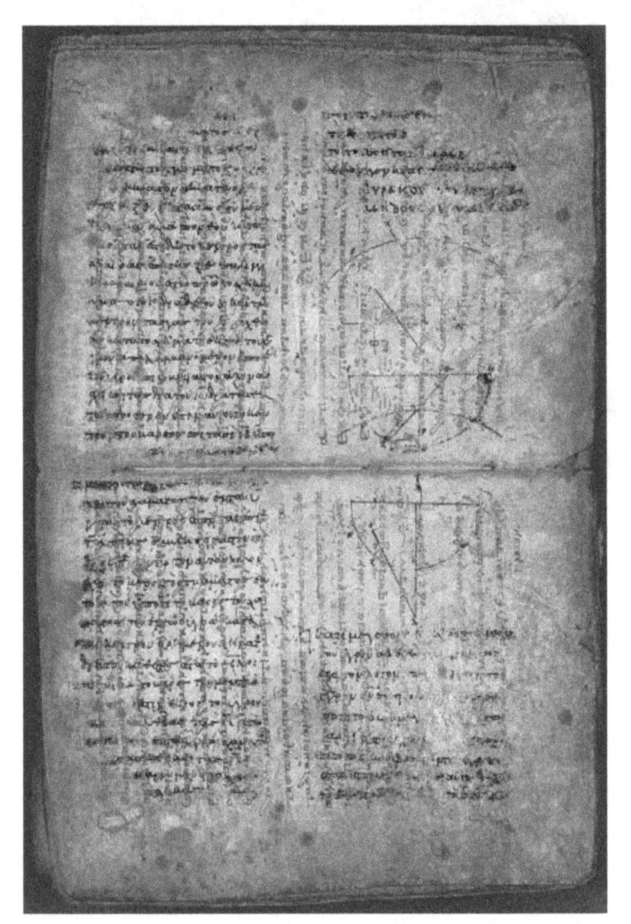

Figure 10.8 The high cost of parchment encouraged the erasure and reuse of entire manuscripts. The Archimedes Palimpsest is a 13th-century Byzantine Greek prayer book made of parchment taken from several earlier manuscripts (which were then partially erased), including most importantly a 10th-century copy of at least seven treatises by the ancient Greek scientist and philosopher Archimedes. [Wikimedia Commons.]

Figure 10.9 This magico-medical scroll of parchment from Ethiopia is now held in London in the Wellcome Collection, Oriental Ms. IX. Its form as a scroll likely dictated storage needs and the way in which the text could be used. [Photo shared under a CC BY 4.0 International License. Wellcome Collection.]

The earliest texts on parchment took the form of rolls or scrolls. For instance, the ancient pharaohs of Egypt (as early as c. 1500 BCE) recorded their law codes on leather rolls. The stitching between multiple pieces of parchment proved much stronger than the pasted joints of papyrus rolls. These scrolls resembled the modern Torah (Hebrew Bible) still featured in Jewish worship, as well as the magical and medical texts recorded in Ge'ez on scrolls in Ethiopia between the 18th and 20th centuries.

Although papyrus manufacture continued for centuries, parchment began to supplant it as the most widely used writing surface in the West by the 4th century, which, as we will see below, was the same period in which the codex definitively replaced the use of scrolls. In fact, during the period in which both parchment and papyrus were employed, scribes copied many works of ancient literature from papyrus to parchment. Had this not occurred, many ancient Greek and Roman classics would not have survived for us today. In the Arab world, the transition from papyrus to parchment occurred somewhat later due to the continued accessibility of the former from Egypt. Parchment was used widely in the Muslim world by the 9th century, but since papermaking technology had arrived from the East in the mid-8th century, demand for parchment in the Muslim world was never as great as in Europe, where paper arrived much later.

Activity: Watch a video

In this BBC video (https://www.youtube.com/watch?v=2-SpLPFaRd0), Wim Visscher gives Dr. Stephen Baxter a modern demonstration of how to make parchment from animal skins.

Organizing Knowledge: The Rise of the Codex

As early as the 1st century CE in the Roman Mediterranean, some papyrus but especially parchment texts began to take a different shape. Rather than retaining the shape of a roll, scribes gathered parchment leaves in a manner similar to wax tablets, which were often bound together as a collection for transcribing longer texts. Thus, the word **codex** originally referred to two or more tablets fastened together.[11]

In the first centuries of the Common Era, scribes began for the first time to employ sheets of parchment in this fashion. They were sewn together at their central crease to form something akin to what we know as a book. Like scrolls, codices had a cover or binding, but in this case they were typically made of wooden boards (which today have been replaced by cardboard). Decorated with cloth, parchment, leather, metal, or precious stones, wooden boards helped identify the content of books. Books might include affixed tags or labels with the name of the author and title of the work on their spines. As we will see below, although not all genres of texts were copied into codices, Christian texts like the Bible, saints' lives, and liturgical works were

Figure 10.10 A Roman fresco from the Praedia of Julia Felix in Pompeii, and now in the Museo Archeologico Nazionale (Naples), displays a bound set of wax tablets and a stylus in the middle section of the lower shelf. It sits next to a contemporary scroll, suggesting that the two writing media were used at a contemporary period. [Praedia of Julia Felix in Pompeii, Naples National Archaeological Museum. Photo by Carole Raddato, shared under a CC BY-SA 2.0 Generic License. Wikimedia Commons.]

frequent candidates for such transitions because of the ways in which scholars studied them.

In China, by contrast, the scroll had a much longer life in Chinese culture but was supplanted in the 10th century by what were known as "whirling books," which opened like an accordion. This design made various parts of the text more accessible than scrolls, which were difficult to manipulate. However, a significant flaw of the "whirling books" was that the paper from which they were fashioned was not very durable along the folded sections. During the Song dynasty (960–1279 CE), "butterfly" bindings became increasingly popular. This format involved printing only on one side of the paper and then folding it so that the blank sides were not visible.[12]

Similar to this technology, but with no direct connection to it, was an accordion-like instrument developed by Amerindians like the Mexica and Maya in Mesoamerica before the time of the Spanish Conquest in the 16th century. Used for recording religious rituals and other significant information, these artifacts were made of a paper-like material composed of pounded tree bark or roots. Like paper, this substance could be polished in preparation for painting symbols and images. Unfortunately, few original examples of this technology survive today since Spanish missionaries condemned indigenous works as idolatrous and confiscated and burned them when they had the opportunity.[13]

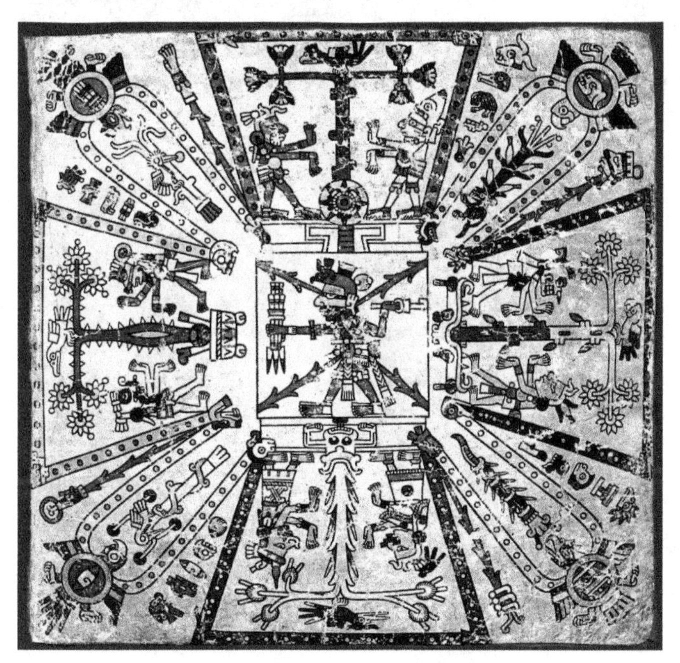

Figure 10.11 The Codex Fejérváry-Mayer is a rare surviving manuscript dating from before European contact (dated to some time between 1200 and 1521 CE). It is made from deerskin parchment and contains the sacred Aztec calendar. [World Museum Liverpool. Wikimedia Commons.]

By contrast, in the European West, scribes fashioned codices from several sheets of parchment cut to the same rectangular size and shape. They typically laid four of these together and folded them once to form the writing unit known as a gathering, which was then sewn at the seam. This gathering produced eight folios (two-sided) or 16 leaves (single-sided), and was called a **quaternio or quire**. So that quires could be bound in the correct sequence, they were marked with signatures (consecutive letters or numbers) on the last page at the top or bottom. From the late 11th century, the first words of the next quire became signature marks. The regular employment of page numbers in a codex did not become common in Europe until the 13th century, and the practice was firmly established by the 16th century.[14]

So what were the advantages of the codex over the scroll? To start, the codex offered convenience to scholars and worshippers, especially with large devotional texts like the Bible in Judaism and Christianity and the Qur'an in Islam. The format of the codex allowed people much greater ease of movement through long and complex texts used in prayer or study. A reader could mark the leaves in a permanent or impermanent way at relevant passages or read pages out of sequence if it was desirable to review a passage or skip ahead. Moreover, as opposed to a scroll, a book could be held in one hand, allowing a scribe to read and write at the same time, something we largely take for granted today. Historians of technology think that the spread of Christianity and Islam helped popularize the effectiveness of the codex over the scroll, not just for sacred texts but also as a medium for organizing and conveying knowledge more effectively.

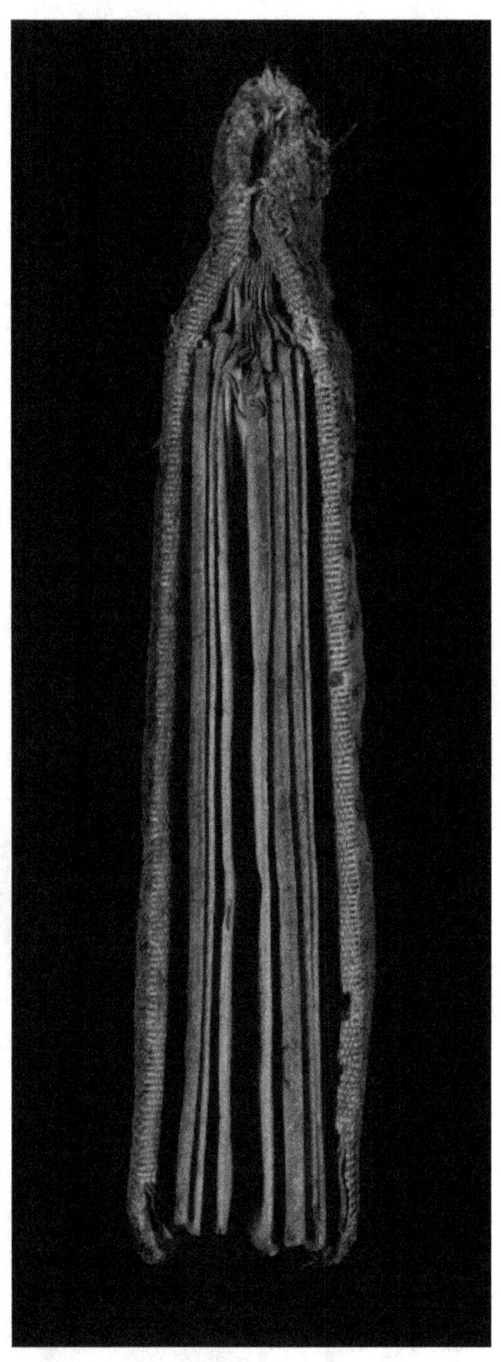

Figure 10.12 This 15th-century folding almanac, Ms. 8932, at the Wellcome Trust was composed in Latin and contains astrological tables and diagrams. Judging from its binding, a physician might have traveled with this luxury manuscript hung from his belt for both practical reasons and as a sign of his status as a learned man. [Wellcome Collection.]

Medieval Manuscript Production and Consumption

In medieval Europe, the main centers for the production, preservation, and copying of texts onto parchment were Christian monasteries. These were filled with monks or nuns, individuals who had pledged to lead a life of chastity, poverty, and obedience. Monastic houses typically had scriptoria, rooms where monks and nuns spent much of their time copying manuscripts by hand. Monastic leaders like Saint Benedict of Nursia in southern Italy in the 6th century emphasized in his *Rule* the importance of the written word in the lives of monks and nuns, and placed an emphasis on public readings of the Bible and other holy texts during mealtimes. Court chanceries were another location where clerics prepared documents during the Middle Ages. From the 11th century, notaries who drafted legal contracts could also be found in the commercial centers of Europe.

In the 11th and 12th centuries, however, book production in Europe shifted in part to cathedral schools and universities and became the source of learning about topics like law, medicine, mathematics, and the humanities. From the 13th century, universities in Italy, France, Spain, Germany, and England (staffed largely by Dominican and Franciscan friars) became the most important locations of textbook production, especially the composition and copying of commentaries and glosses on canonical authorities.

Have you ever wondered how medieval university students procured the books they needed for their studies? Medieval students (all male institutions—women were not allowed to participate) typically rented texts through local booksellers through what was known as the *pecia* system. Namely, rather than buying books, students rented exemplars by the **quire** (known as a *pecia*), and used them to make their own copies. Once returned, booksellers rented the same quires to other students. This system had the advantage of decreasing the replication of errors since each copy depended upon the same uncorrupted text. Students often sold their own copies when their courses were finished in order to pay off debts, including their beer tabs at the local bars so common in university towns.[15]

Just because manuscripts were the bread and butter of medieval university learning does not mean that they were accessible to anyone but a small number of men, many of them clerics, who received a higher education. Scholastic manuscripts were written in Latin, which was no longer commonly spoken in Europe. They thus required significant learning to copy accurately and to comprehend. Even worse, they included extensive use of abbreviations and there were often no distinct separations between words. Although these conventions allowed students to copy quickly, readers had to read the texts aloud, a practice that only changed in the 16th century when efforts were made to make texts more legible. These traditions and the use of Latin excluded many potential readers from access to these texts.

What kinds of texts were popular among the medieval elites who could afford them (or were members of institutions like monasteries that owned them)? The most common manuscripts were Latin bibles and books of hours, the latter being prayer books for private devotions, many of which were used by women. In the later Middle Ages, the most commonly copied manuscripts included annals, chronicles, and romances. Many of these were composed in the vernacular rather than Latin, which by this point was mainly limited to clerics. By the 13th century, wealthy patrons, both male and female, who wanted to expand their private libraries commissioned manuscripts from booksellers in cities like Paris. These shops or *scriptoria* employed scribes and illuminators to create custom copies.[16]

Figure 10.13 Medieval books of hours such as this one associated with the 15th-century female aristocrat Catherine of Cleves encouraged pious contemplation. The mouth of hell depicted on the left was meant to instill fear of sin in its female reader. This work illustrates how books served both devotional purposes as well as exhibiting the wealth of their owners. ["Office of the Dead," MS M.917/945, ff. 168v–169r, Morgan Library and Museum.]

Plant Fiber Technology: The Advent of Paper

Paper was invented in China in the second century BCE. Composed of vegetal fibers like hemp, paper's cost was modest and it quickly replaced silk fabric as a writing material for all but luxury manuscripts. By the 2nd century CE, paper goods probably entered into trade through the intricate network of roads and tracks that crossed Eurasia: the routes between China and the West known as the Silk Road. Travel across the Eurasian landmass was rudimentary and expensive with the aid of mules, camels, and oxen; the overland itinerary also depended upon the political stability of the regions through which tradesmen passed.[17]

Knowledge of paper and the process by which it was made from cloth rags (usually hemp or linen) reached Persia by the 7th century and the city of Samarqand by the 8th century. From there, the use of paper spread throughout the Islamic world in the 9th century, promoted by the Abbasid caliphs who employed it for official records. In Spain, which was controlled by Muslim rulers from the early 8th century, the know-how for paper-making passed to the Christian West. Eleventh-century Valencia, for instance, was an important urban center for paper manufacturing.[18] Italian papermakers refined these methods in Fabriano, including inventing the watermark, adopting gelatin sizing, and improving pulping methods. By the 14th century, Italian paper made in paper mills dominated European markets.

As opposed to parchment, paper provided a comparatively cheap writing surface. In the era before mechanical printing, we know that paper profoundly influenced scribal culture in medieval Islamic lands, since it was more abundant and accessible than parchment. However, paper was also more easily adapted to woodblock printing popular in China and Korea from at least the 8th century. In Western Europe, paper's future looked very bright following the development of moveable type in the mid-15th century. For this reason, paper largely supplanted parchment by the 16th century due to cost, accessibility, and fit with new printing technologies. Paper thus had the important effect of opening up literate Christian culture to a much wider (though certainly not universal) array of readers than could afford such works when parchment was the prevalent writing material.

Our knowledge of this history has important lessons that we can bring to bear on our understanding of the social impact and accessibility of digital media. Namely, how is the transition to digital technology affecting how we compose literature, organize and present information, and store data? While our ever-increasing dependence on computers, tablets, and smartphones may allow us to marshal large amounts of data from a seemingly endless number of sources, we also need to ask how individuals and groups benefit from this information and whether all benefit equally from the digital revolution.

The Invention of Printing

Efforts to make exactly reproducible multiple impressions go back thousands of years. In China, stone steles or upright stone slabs with engraved calligraphic texts were used to make rubbings or copies. As we have seen, in ancient Mesopotamia, stone cylinder seals were engraved around their perimeter so that when they were rolled across a receptive surface, they revealed their messages. The oldest printed material known (dated to 751 CE) is from Korea, and is now called the **Dharani Sutra**. The earliest extant dated printed work, the Chinese Diamond Sutra scroll, was printed from a woodblock in 868. The carving of over 80,000 woodblocks between 1236 and 1251, many of which still survive, supported an effort to print the Korean Buddhist scriptures in their entirety. In China, Pi Cheng innovated with movable type made from fire-hardened clay or liquid glue under the Sung Dynasty (960–1279). In Korea, cast metal type was first developed in 1234, two centuries before it made its debut in Western Europe.[19]

Figure 10.14 This detail comes from the woodblock print on the frontispiece of the Diamond Sutra, found in Cave 17 at Dunhuang in China. It is the oldest extant printed book with a firmly known date of 868 CE. [British Library.]

In the mid-15th century, the German goldsmith Johannes Gutenberg, among others, invented cast-metal movable type in individual characters designed to be printed with a press. Once a page of type was prepared, printers inked and printed them onto dampened paper with a wooden press adapted from those traditionally used to make wine or oil. The pages were hung out to dry and then proofread and gathered in the correct order. Printers also carved historiated or decorated initials and other illustrations onto wooden blocks made at the same thickness as the type so that they could be set in the form. They included this labor-intensive step to preserve a style that was common to hand-copied manuscripts on parchment and paper. Efforts to imitate the style of hand-copied manuscripts suggest that consumers of printed books considered the latter desirable. In fact, for a century after the advent of printing, luxury copies of the Bible were still copied by hand and printed copies in many cases were made to look like those written by hand.[20]

Activity: Watch Gutenberg-style printing

This video (https://www.youtube.com/watch?v=DLctAw4JZXE) demonstrates the procedure of printing a single leaf on a Gutenberg-style press.

Gutenberg first published the Bible in 1456 using metal type, a wooden press, and Italian paper. Although he did make some luxury copies for elite clients on parchment, most printing was done on paper. From Germany, this important invention spread rapidly to workshops in Switzerland, Italy, Spain, France, and England. It created considerable challenges for those who wished to print texts in non-Roman alphabets such as Arabic, in which letter shapes varied depending upon their position at the start, middle, or end of a word, or Chinese, in which there are thousands of characters. Such considerations remain important today for hardware engineers designing computer keyboards and touchpads for global markets.

We should keep in mind that many kinds of documents like wills and private letters were not affected by the advent of printing press for centuries. Moreover, printing aggravated instead of eliminated some of the shortcomings of hand-copied manuscripts since rather than creating one defective copy, one might produce hundreds of corrupted copies. For this reason, for centuries after Gutenberg's invention, distinct communities remained loyal to the tried and true method of copying texts by hand: this was especially true of monks and nuns since it was a required component of the monastic day. Moreover, the creation of luxury manuscripts was a business controlled by well-organized and powerful guilds, especially in university towns where students were a source of profit. In the early decades of the print trade, guilds protected their interests by running competing printing workshops out of town.

The Printed Word: Access, Literacy, and the Publishing Industry

What were the ramifications of the invention of printing with moveable type in the West? This is an enormous subject, but here is some food for thought.[21]

To start, the set-up cost involved in printing was quite prohibitive. For instance, in Italy in 1483, the Ripoli Press charged three florins to set up and print a quinterno (quire) of the Italian priest and scholar Marsilio Ficino's translation of Plato's *Dialogues*. This price might seem quite expensive if we know that a contemporary scribe might have charged a single florin per quinterno to duplicate the same work. However, the Ripoli Press produced 1,025 copies of this work whereas the scribe could make just one at a time. Printing thus had the advantage of producing multiple copies at a reduced rate per copy. In other words, there were significant economies of scale for works like the Bible for which there was always demand or for pamphlets that required broad circulation. This situation meant that sizeable libraries were no longer the exclusive preserve of monasteries or the elite.

It is certainly true that printing made many texts much more available, since bookshops could now print books on their lists in large quantities and libraries could acquire more works at a higher rate of speed. The end result was that scholars and literate members of the public had access to a much broader range of texts than had previously been possible. These workshops thus opened up the possibility of owning a book and stimulated interest in literature (and literacy) among a larger public readership than was the case previously.

An additional outcome of mechanical printing was the commercialization of the reproduction of books, which some scholars have convincingly argued contributed to the commodification of the book. In other works, if books were increasingly seen as a desirable possession that was in reach of a greater number of consumers, printers competed for this market by making their books more attractive in a variety of ways. These options affected which works booksellers chose for printing, how they embellished their covers, and the choices they made about decorative fonts or large numbers of illustrations.

Printing and Revolutionary Thinking

The first print shops were centers of intellectual and commercial collaboration in which editors, proofreaders, writers, merchants, and patrons of learning came together to produce books collaboratively. A useful corollary of that development was the serendipitous juxtaposition of texts and maps in early printing workshops. In other words, when everything from Bibles to atlases to works of science were brought together under a single roof, they supported non-traditional thinking and afforded opportunities for innovation that might not have otherwise been conceivable. An analogous example in the modern world is our daily dependence on digital search engines like Google that allow us to input a keyword regardless of topic. This kind of search often produces very different results than if we followed the more old-fashioned method of searching by name of author or in a single subject area. While such finds are useful for scholarly research only if they are put in their proper context, they nonetheless widen the spectrum of possibilities as was the case for those who frequented early modern printing workshops.

In a time when we are all familiar with the mantra of the internet bringing about a more democratic world, a promise that has proved somewhat hollow and elusive at best, thinking about the invention of printing has particular resonance. The advent of mechanical printing had a number of unintended consequences by legitimizing and popularizing works that would not have previously circulated widely because they were seen as suspect by the Catholic Church. To be certain, religious leaders were concerned about the distribution of unauthorized works on theology, astrology, alchemy, and magic.

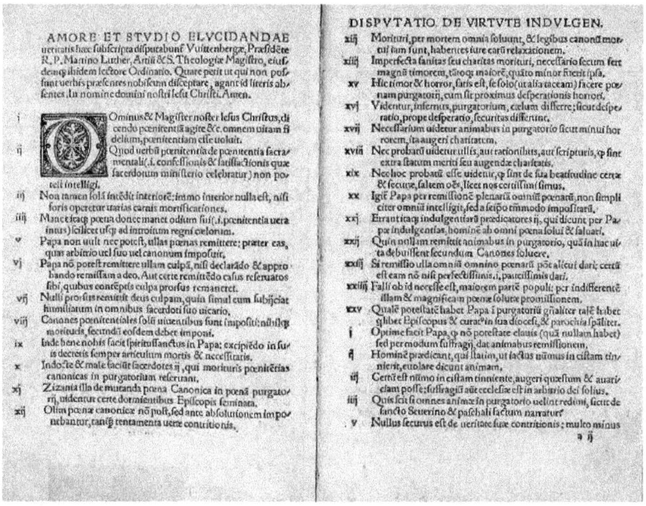
AMORE ET STVDIO ELVCIDANDAE ueritatis hæc subscripta disputabunt Vuittenbergæ, Præsidēte R. P. Martino Luther, Artiū & S. Theologiæ Magistro, eiusdemq; ibidem lectore Ordinario. Quare petit ut qui non possunt uerbis præsentes nobiscum disceptare, agant id literis absentes. In nomine domini nostri Iesu Christi. Amen.

i Dominus & Magister noster Iesus Christus, dicendo pœnitentiā agite &c. omnem uitam fidelium, pœnitentiam esse uoluit.

ij Quod uerbū pœnitentia de pœnitentia sacramentali (.i. confessionis & satisfactionis quæ sacerdotum ministerio celebratur) non potest intelligi.

iij Non tamen solā intēdit interiorē; immo interior nulla est, nisi foris operetur uarias carnis mortificationes.

iiij Manet itaq; pœna donec manet odium sui (.i. pœnitentia uera intus) scilicet usq; ad introitum regni cælorum.

v Papa non uult nec potest, ullas pœnas remittere; præter eas, quas arbitrio uel suo uel canonum imposuit.

vj Papa nō potest remittere ullam culpā, nisi declarādo & approbando remissam a deo. Aut certe remittēdo casus reseruatos sibi, quibus contēptis culpa prorsus remaneret.

vij Nulli prorsus remittit deus culpam, quin simul eum subijciat humiliatum in omnibus sacerdoti suo uicario.

viij Canones pœnitentiales solū uiuentibus sunt impositi; nihilq; morituris, secundū eosdem debet imponi.

ix Inde bene nobis facit spiritussanctus in Papa: excipiēdo in suis decretis semper articulum mortis & necessitatis.

x Indocte & male faciūt sacerdotes ij, qui morituris pœnitētias canonicas in purgatorium reseruant.

xj Zizania illa de mutanda pœna Canonica in pœnā purgatorij, uidentur certe dormientibus Episcopis seminata.

xij Olim pœnæ canonicæ nō post, sed ante absolutionem imponebantur, tanq; tentamenta ueræ contritionis.

DISPVTATIO DE VIRTVTE INDVLGEN.

xiij Morituri, per mortem omnia soluunt, & legibus canonū mortui iam sunt, habentes iure earū relaxationem.

xiiij Imperfecta sanitas seu charitas morituri, necessario secum fert magnū timorem, tantoq; maiorē, quāto minor fuerit ipsa.

xv Hic timor & horror, satis est, se solo (ut alia taceam) facere pœnam purgatorij, cum sit proximus desperationis horrori.

xvj Videntur, infernus, purgatorium, cælum differre; sicut desperatio, prope desperatio, securitas differunt.

xvij Necessarium uidetur animabus in purgatorio sicut minui horrorem, ita augeri charitatem.

xviij Nec probatū uidetur ullis, aut rationibus, aut scripturis, q sint extra statum meriti seu augendæ charitatis.

xix Nec hoc probatū esse uidetur, q sint de sua beatitudine certæ & securæ, saltem oēs, licet nos certissimi simus.

xx Igit Papa per remissionē plenariā omniū pœnarū, non simpliciter omniū intelligit, sed a seipso tm̄modo impositarū.

xxj Errant itaq; indulgentiarū prædicatores ij, qui dicunt per Papæ indulgentias, hominē ab omni pœna solui & saluari.

xxij Quin nullam remittit animabus in purgatorio, quā in hac uita debuissent secundum Canones soluere.

xxiij Si remissio ulla omniū omnino pœnarū pōt alicui dari; certū est eam nō nisi perfectissimis, .i. paucissimis dari.

xxiiij Falli ob id necesse est, maiorem partē populi: per indifferentē illam & magnificam pœnæ solutæ promissionem.

xxv Qualē potestatē habet Papa i purgatoriū gn̄aliter talē habet qlibet Episcopus & curat⁹ in sua diocesi, & parochia spāliter.

j Optime facit Papa, q nō potestate clauis (quā nullam habet) sed per modum suffragij, dat animabus remissionem.

ij Hominē prædicant, qui statim, ut iactus nūmus in cistam tinnierit, euolare dicunt animam.

iij Certū est nūmo in cistam tinniente, augeri quæstum & auariciam posse; suffragiū aūt ecclesiæ est in arbitrio dei solius.

iiij Quis scit si omnes animæ in purgatorio uelint redimi, sicut de sancto Seuerino & paschali factum narratur?

v Nullus securus est de ueritate suæ contritionis; multo minus

a ij

Figure 10.15 In 1517, the then-Catholic monk Martin Luther is thought to have posted his 95 Theses on the door of the church in Wittenberg, Germany, to protest what he saw as corrupt practices including the sale of indulgences for salvation. Although the document was originally handwritten in Latin, by the following year he translated it to German and printed it. This technology allowed him to circulate numerous copies of the document in a short period of time. Church authorities were not able to stop the spread of what they saw as dangerous ideas. [University of Basel Library.]

Religious authorities were not the only public figures made uncomfortable by the freedom of expression offered by printing. Powerful lay rulers like monarchs and aristocrats were especially concerned about individuals and groups that sought to reach a public audience with pamphlets and books criticizing secular leaders and advocating alternative political models. The relative ease with which an individual could reproduce hundreds of copies of any particular work meant that it was difficult for monarchs to enforce effective bans on texts they deemed unacceptable.

Martin Luther's translation of the Bible from Latin to German in the early 16th century challenged the longstanding monopoly of the Catholic Church. Unlike his predecessors, whose manuscripts were burned when they were condemned and executed as heretics, Luther circulated thousands of copies far and wide in the Holy Roman Empire. Luther's successful resistance to the Church's demands and his evasion of Church censors led to the birth of the breakaway religious movement of Lutheranism. Scholars suspect that Luther would not have succeeded in his reform agenda had it not been for the technological advantage of printing.

Has Digitization Changed Our Relationship with the Written Word?

Why is it important to learn about the birth of the codex? The French historian Roger Chartier has suggested that we consider the digital revolution in terms of the "longue durée" or the long term. In other words, we need to think about the possibilities digital texts and their transmission create not just in our lives today but also over the coming decades and centuries.[22]

First, how does the digital revolution change how we engage with a text? Theoretically, at least, readers can interact with texts more fluidly. Previously, we underlined, took notes, or wrote in the margins of books. Today, readers can index, annotate, copy, recompose, hyperlink, and move digital texts. In essence, each person becomes an active participant in the text. In this way, digital works blur the distinction between reading and writing and between author and reader, since any reader with a digital device has the ability to create new texts from fragments that have been spliced and reassembled. Of course, there is a thin line between this practice and plagiarism if one claims this work as his or her own without acknowledging the original sources of such blended works. If we look around the classroom today, however, not all students have abandoned paper and many continue to take notes the old-fashioned way. Indeed, recent studies suggest that the act of writing (as opposed to keying words into a computer or tablet) helps us to process and retain information more effectively.[23]

Second, there is no doubt that the digital revolution has changed how we order and store information. With magnetic storage devices, readers can construct unique collections of original texts whose existence and organization depend upon their individual whims (and the texts to which they can gain access). One can not only store but also modify or rewrite these works at any moment. These technologies challenge not only traditional ideas of literary property and copyright, but also reshape our conceptions of what a library is, what one should do there, and how it should look. We can each create our own libraries in a way that was not long ago confined exclusively to institutions and elites. However, we must also be conscious that these libraries are easily erased or destroyed. All of us are familiar with the panic that sets in if our computer crashes or our hard drive fails. With rapidly changing technology, storage technology like floppy disks have quickly gone out of use as they were replaced by CDs, DVDs, and USB drives. Not only are older formats difficult or impossible to read with a new computer, but these storage devices were never intended for the long term. Most forms of magnetic storage must be backed up since they become unreadable in less than a decade.

Figure 10.16 A photo of the fourth-floor stacks of the Free Library of Philadelphia, a city that received one of the largest grants offered by Andrew Carnegie. With these funds, Philadelphia constructed 25 branch libraries between 1905 and 1930. Although the Central Library, which houses the Free Library chartered in 1891, was not part of the endowment, it became one of the most important libraries in the world with over one million volumes. [Photo by Joseph Elliott, Historical American Buildings Survey, Library of Congress.]

These many factors suggest that the digital revolution will not cause the printed word to disappear overnight as industry pundits often predict. A historical perspective suggests that just as the scroll imposed its organization on the early codex, so handwritten codices imposed their form upon the earliest printed texts. These shaped the early form, structure, and layout of the successive media in which knowledge was conveyed. Essentially, one could argue that printed books have done the same to ebooks, especially in light of users' attachment to the tactile experience they love about reading an authentic paper book or newspaper. Over the course of several generations, however, it is likely that the memory of this experience will fade and new readers will prioritize some aspects of book culture while eliminating others depending upon their needs. With that, some features of the codex will remain embedded while others will be discarded.

Information Storage Technologies as Appropriate Technologies

Looking at the history of writing technologies teaches us that humans use different materials for storing information in different social contexts. Our choice of a material for writing, for example, depends upon the availability and accessibility of materials, the needs of users in various places and times, and established local cultural practices. Papyrus worked well where it could be locally cultivated and was used for texts that were not expected to be preserved for long periods of time. Parchment, by contrast, worked well for large texts that needed to be consulted for specific information rather than necessarily reading it from front to back. And parchment was effective for storing important information that was meant to be studied and guarded for generations, especially in regions of the world that were damp and cold. Neither of these writing technologies, however, supplanted the use of stone inscriptions which continued to be used for gravestones, ceremonial markers, and engraved sculpture. Similarly, today, stone, paper, and hard drives coexist as materials for information storage because they each work in a particular constellation of practice.

> Think about why you might find it easier to jot a library reference number on a piece of scrap paper if you did not want to bring along a mobile phone, tablet, or computer into the library stacks to search for a book. Or why you might write someone's phone number on your hand when making new friends at the pool because you did not bring your own cell phone (no pockets). By contrast, consider why the ubiquitous phone books and "yellow pages" that used to be in every household and business have now disappeared, since internet searches and online advertising have proved to be more effective ways to convey this information and have thus made these bulky storage devices obsolete.

The idea that a material technology is relevant to its local context of use is termed **appropriate technology**. Put simply, a material has to work easily and well in people's everyday lives. It must be appropriate to their needs and compatible with local social organization, but it also must be affordable, locally sourced or available, locally repairable (if intended to last some time), and work symbiotically with the local material environment.[24] A digital storage center deployed in an area without a stable electricity grid will be just as ineffective as requiring royal and church officials to write on papyrus in 16th-century Northern Europe. For a recent example, we might think about how the COVID-19 pandemic pushed shoppers as never before to abandon printed money in favor of contactless transactions. However, this change does not mean that printed money will disappear any time soon since it is still considered to have value beyond the immediate crisis, and the costs associated with this technology (and the oversight embedded in it) may not be feasible or desirable in all cases. The entanglements of writing materials, and the various implications of their use, mean that different writing materials necessarily coexist. And, our use of one material may influence how we deploy another.

For materials scientists and engineers, the concept of appropriate technology can focus development teams on creating technologies with new materials that are sustainable in the context of their use, versus developing technologies that work in one culture (where, for instance, electricity is dependable) but are inefficient, unaffordable, or simply alien to another culture of use (where power might be available one hour per day or users are concerned about government oversight). However, just as a culture shapes our use of materials, the introduction of new materials for information storage shapes our writing and preservation practices. Thinking of materials as appropriate reminds us to use new digital storage materials purposefully, bearing in mind what they help us to do best, and what other writing materials are still needed to make and share information in a given society.

We may close with Chartier's warning that: "The library of the future must also be a place that will preserve the knowledge and understanding of written culture in the forms that were and still are today, very much its own. The electronic representation of all texts whose existence did not begin with computerization should not in any way imply the relegation, forgetting, or, worse yet, destruction of the objects in which they were originally embodied. More than ever, perhaps, one of the critical tasks of the great libraries is to collect, to protect, to inventory. . ."[25]

The message that we should take from this warning are the potential challenges of digitization. We must evaluate what digital formats and electronic storage mean for the quality of our lives, engagement with knowledge, and preservation of information for the near and long-term future. While change may be inevitable, and a current generation of children will never know life in the absence of these technologies, we must observe how they affect our day-to-day existence, the ways in which we learn, and the methods by which we preserve knowledge of the past. These choices will indelibly shape how humans live together in the future.

Discussion Questions

1. Condorcet believed that the limitless reproduction of texts afforded by moveable print was key to the success of democracy, and he credited this technology with greater access to justice in human society. Do you think

our current on-line digital technology increases access to justice?
2. In what ways has digital storage changed what we study and how we learn?
3. How has today's technology affected communication, positively and negatively?
4. What is the property of great importance to remember when deciding whether to use magnetic information storage or some other form of data storage?
5. What materials have been used throughout history to preserve the printed word? Discuss the properties for which they were chosen as a vehicle for the written picture or word.

Key Terms

appropriate technology
codex
Dharani Sutra
ostraca
quaternio or quire
"vulgar writing"

Author Biography

Bonnie Effros (Ph.D., European Medieval History, UCLA, 1994), holds the Chaddock Chair in Economic and Social History in the School of Histories, Languages, and Cultures at the University of Liverpool. Prof. Effros was previously the inaugural Rothman Chair and director of the Center for the Humanities and the Public Sphere in addition to being a Professor in the Department of History at the University of Florida in Gainesville. She is the editor of the Brill Series on the Early Middle Ages and serves on the Editorial Board of *Studies in Late Antiquity*.

Further Reading

Baker, Keith Michael. *Condorcet: From Natural Philosophy to Social Mathematics*. Chicago: Univ. of Chicago Press, 1975. https://archive.org/condorcetfromnat0000bake.

Crick, Julia, and Alexandra Walsham, eds. *The Uses of Script and Print, 1300–1700*. Cambridge: Cambridge Univ. Press, 2004. http://www.worldcat.org/oclc/433605414.

Danesi, Marcel. *Vico, Metaphor, and the Origin of Language*. Bloomington, Indiana: Indiana Univ. Press, 1993. http://www.worldcat.org/oclc/802512895.

Daniels, Peter T. and William Bright, eds. *The World's Writing Systems*. Oxford: Oxford Univ. Press, 1996. http://www.worldcat.org/oclc/909698730.

Diringer, David. *The Book before Printing: Ancient, Medieval and Oriental*. New York: Dover Publications, 1982. http://www.worldcat.org/oclc/611226929.

Acknowledgments

I would like to dedicate this essay to my doctoral advisor Professor Richard H. Rouse, who instilled in me, among other things, an abiding appreciation for the mechanics of books and book culture.

Notes

1. Elizabeth Hill Boone, "Aztec Pictorial Histories: Records without Words," in *Writing Without Words*, eds. Elizabeth Hill Boone and Walter D. Mignolo (Durham: Duke Univ. Press, 1994), 50–76, http://www.worldcat.org/oclc/837771541.
2. Chartier focuses on Vico and Condorcet as symptomatic of these larger questions.; Roger Chartier, "The Representation of the Written Word," in *Forms and Meanings: Texts, Performances, and Audiences from Codex to Computer* (Philadelphia: Univ. of Pennsylvania Press, 1995), 6–24, http://www.worldcat.org/oclc/44961557.
3. Giambattista Vico, *On the Most Ancient Wisdom of the Italians. Drawn out for the Origins of the Latin Language* [1711], trans. Jason Taylor (New Haven: Yale Univ. Press, 2010), http://www.worldcat.org/oclc/601348202.
4. Condorcet, "Sketch for a Historical Picture of the Progress of the Human Mind: Tenth Epoch," trans. Keith Michael Baker, *Daedalus* 133, no. 3 (Summer, 2004), 65-82, https://www.jstor.org/stable/20027931.
5. Jean-Jacques Glassne, *The Invention of Cuneiform: Writing in Sumer* (Baltimore: Johns Hopkins Univ. Press, 2003), http://www.worldcat.org/oclc/51041422.
6. Roderick Whitfield and Anne Farrer, *Caves of the Thousand Buddhas: Chinese Art from the Silk Route* (New York: George Braziller, 1990), http://www.worldcat.org/oclc/20931666.
7. Bridget Leach and William John Tait, "Papyrus," in *Ancient Egyptian Materials and Technology*, eds. Paul T. Nicholson and Ian Shaw (Cambridge: Cambridge Univ. Press, 2000), 227–53, http://www.worldcat.org/oclc/1087529499.
8. Michael McCormick, *Origins of the European Economy: Communications and Commerce, AD 300–900* (Cambridge: Cambridge Univ. Press, 2001), 696–728, http://www.worldcat.org/oclc/469934775.

9. Kelly Sloane, "Epic Illustrations: Vergil's Aeneid in the Vergilius Vaticanus," 2005-2006 Penn Humanities Forum on Word & Image, Undergraduate Mellon Research Fellows, https://web.archive.org/web/20150312073624/http://humanities.sas.upenn.edu/05-06/mellon_uhf.shtml.
10. Ronald Reed, *The Nature and Making of Parchment* (Leeds: Elmete, 1975), http://www.worldcat.org/oclc/2038984.
11. J. A. Szirmai, "Wooden Writing Tablets and the Birth of the Codex," *Gazette du Livre Médèvale* 17 (1990): 31–32, https://www.persee.fr/doc/galim_0753-5015_1990_num_17_1_1144.
12. Tsien Tsuen-Hsuin, "Paper and Printing," in *Science and Civilization in China*, Vol. 5, sect. 1, ed. Joseph Needham (Cambridge: Cambridge Univ. Press, 1985), 227–33, http://www.worldcat.org/oclc/79441618.
13. Alan R. Sandstrom and Pamela Effrein Sandstrom, *Traditional Papermaking and Paper Cult Figures of Mexico* (Norman: Univ. of Oklahoma Press, 1986), 3–31, http://www.worldcat.org/oclc/12950001; Walter D. Mignolo, "Signs and Their Transmission: The Question of the Book in the New World," in *Writing Without Words*, 220–25.
14. Bernhard Bischoff, *Latin Palaeography: Antiquity and the Middle Ages*, trans. Dáibhí Ó Cróinín and David Ganz (Cambridge: Cambridge Univ. Press, 1990), 20–48, http://www.worldcat.org/oclc/924973935.
15. Mary A. Rouse and Richard H. Rouse, "The Book Trade at the University of Paris, ca. 1250–ca. 1350," in *Authentic Witnesses: Approaches to Medieval Texts and Manuscripts* (Notre Dame: Univ. of Notre Dame, 1993), 259–338, http://www.worldcat.org/oclc/925020363.
16. Christopher de Hamel, *A History of Illuminated Manuscripts*, 2nd ed. (London: Phaidon Press, 1997), http://www.worldcat.org/oclc/611946900.
17. Tsuen-Hsuin, "Paper and Printing," 52–83.
18. Helen Loveday, *Islamic paper: a study of the ancient craft* (London: Archetype Publications, 2001), http://www.worldcat.org/oclc/249681685.
19. Jixing Pan, "On the Origin of Printing in the Light of New Archaeological Discoveries," in *Chinese Science Bulletin* 42.12 (1997): 976–81, https://doi.org/10.1007/BF02882611.
20. Elizabeth L. Eisenstein, *The Printing Revolution in Early Modern Europe*, 2nd, rev. ed. (Cambridge: Cambridge Univ. Press, 2005), http://www.worldcat.org/oclc/8157450657.
21. Lucien Febvre and Henri-Jean Martin, *The Coming of the Book: The Impact of Printing* 1450–1800 (London: Verso, 1997), http://www.worldcat.org/oclc/931383419.
22. Chartier, "The Representation of the Written Word," 6–24.
23. Pam A. Muelle and Daniel M. Oppenheimer, "The Pen Is Mightier Than the Keyboard: Advantages of Longhand Over Laptop Note Taking," in *Psychological Science* 25, no. 6 (2014): 1159–68, https://doi.org/10.1177%2F0956797614524581.
24. Marianne de Laet and Annemarie Mol, "The Zimbabwe Bush Pump: Mechanics of a Fluid Technology," in *Social Studies of Science* 30, no. 2 (2000): 225–63, https://doi.org/10.1177%2F030631200030002002.
25. Chartier, "The Representation of the Written Word," 24.

CPSIA information can be obtained
at www.ICGtesting.com
Printed in the USA
BVHW051936240322
632297BV00004B/23